Sustainability of Organic Farming in Nepal

Mrinila Singh • Keshav Lall Maharjan

Sustainability of Organic Farming in Nepal

 Springer

Mrinila Singh
Graduate School for International
  Development and Cooperation
Hiroshima University
Higashi-Hiroshima, Hiroshima
Japan

Keshav Lall Maharjan
Graduate School for International
  Development and Cooperation
Hiroshima University
Higashi-Hiroshima, Hiroshima
Japan

ISBN 978-981-10-5618-5   ISBN 978-981-10-5619-2 (eBook)
DOI 10.1007/978-981-10-5619-2

Library of Congress Control Number: 2017952943

© Springer Nature Singapore Pte Ltd. 2017
This work is subject to copyright. All rights are reserved by the Publisher, whether the whole or part of the material is concerned, specifically the rights of translation, reprinting, reuse of illustrations, recitation, broadcasting, reproduction on microfilms or in any other physical way, and transmission or information storage and retrieval, electronic adaptation, computer software, or by similar or dissimilar methodology now known or hereafter developed.
The use of general descriptive names, registered names, trademarks, service marks, etc. in this publication does not imply, even in the absence of a specific statement, that such names are exempt from the relevant protective laws and regulations and therefore free for general use.
The publisher, the authors and the editors are safe to assume that the advice and information in this book are believed to be true and accurate at the date of publication. Neither the publisher nor the authors or the editors give a warranty, express or implied, with respect to the material contained herein or for any errors or omissions that may have been made. The publisher remains neutral with regard to jurisdictional claims in published maps and institutional affiliations.

Printed on acid-free paper

This Springer imprint is published by Springer Nature
The registered company is Springer Nature Singapore Pte Ltd.
The registered company address is: 152 Beach Road, #21-01/04 Gateway East, Singapore 189721, Singapore

Preface

The term "sustainability" entails "sustenance, survival or flourishing of a process, an organism or a resource." Since the beginning of the twenty-first century, it has been talked about in almost every sphere of life – from natural environment to far more complex entities such as human societies, cultural traditions, or social institutions. In the case of farming, this concept came into play when the principle of green revolution could no longer improve situation of food insecurity, resource degradation, and regional imbalance of benefits. Green revolution, which commonly is also known as conventional farming, has clearly benefited through increased crop production with the use of chemical fertilizers and pesticides and high-yielding crop varieties and contributed significantly in reducing world hunger. But overtime we also came to realize that excessive and inappropriate use of chemical fertilizers and pesticides polluted groundwater, streams, rivers, and oceans; triggered land degradation through soil erosion; and severed deterioration of the arable soil. It caused professional hazard; killed beneficial insects and other wildlife; reduced biodiversity; increased pest adaptation and resistance, desertification, and water eutrophication; and affected those who consumed it through food residue. This triggered the search for a more sustainable way of farming, the one that maintains the quantity and quality of agricultural produce over a considerably longer period of time without an exhaustion of ecosystem services that crops and animals depend on for their productivity.

Organic farming is known to be such sustainable practice that provides better working environment for farmers and safe food for consumers and maintains environmental integrity. However, it is often criticized for lower production and difficulty in accessing the premium market to compensate for this loss. The nature and feasibility of organic or conventional farming systems is very context-specific. For example, in rural areas of developing countries, both farming systems integrate crops and livestock, unlike in industrialized countries where conventional farmers are known to have mono-cropping system and almost solely rely on chemical inputs. Similarly, organic farming cannot be labeled exclusively to have lower production as in some areas it has actually resulted in better production than its conventional counterpart. Supporting factors such as infrastructure, irrigation, suitable policies, and so on also play a major role when defining success of any farming system. Moreover, sustainability should be defined from three widely accepted dimensions: social, economic, and environmental. Therefore, to compare the achievability of these dimensions among organic and conventional farming systems and to understand the degree of variation between these two farming systems, this book did a case study of Chitwan District in Nepal where indiscriminate use of agrochemicals is quite common but group conversion to organic farming also exists in three adjoining village development committees: Phoolbari, Shivanagar, and Mangalpur. About 285 respondents were selected by stratifying individual households based on whether or not they are member in an organic farming-related group. Data were analyzed both descriptively and utilizing appropriate econometric tools.

This book is comprised of twelve chapters. Chapter 1 gives an overview of sustainability of a farming system by discussing the social, economic, and environmental aspect of organic farming system in relation to conventional farming system. It also highlights how eminent organic farming has gotten over the years world over and discusses its status in South Asia, a region distinguished for being economically backward. After stating the regional status of organic farming, Chap. 2 discusses the state and scope of organic farming in Nepal. Organic farming holds huge potential in Nepal due to exclusion of costly agrochemicals, ecological diversities, and higher labor availability in farming sector. Although Nepalese government is mostly inclined toward supporting conventional farming system, its support for organic farming has gradually increased in recent years through its policies. However, lack of adequate and integrated research, extension, manpower, and other support on organic farming production, marketing, and input supply has hindered the development of organic farming promotion. Chapter 3 provides glimpse of nature of organic farming in Chitwan District with special focus on a group formed for the purpose of organic farming. Group formation has led to higher adoption rate of organic farming, although a closer scrutiny shows that being a member of such group does not guarantee that all farmers will undeniably practice organic farming over the years. It discusses differences in households practicing organic and conventional farming systems in terms of their socioeconomic characteristics and the functioning of the group itself that has led some farmers to divert their practice back to conventional farming.

Chapter 4 is related to social dimension of a farming system where both organic and conventional farmers are analyzed based on their socioeconomic characteristics. More than group membership, households receiving higher number of organic farming-related training have highly significant probability to continue practicing organic farming. The impact of longevity of group formation and vitality of training is also reflected in higher adoption rate of organic farming across different village development committees considered for this study. Likewise, commercially available organic fertilizers and pesticides also seem to be playing a significant role in the adoption rate of organic farming.

Chapters 5, 6, and 7 discusses environmental dimension of a farming system. Chapter 5 evaluates differences in soil properties. No significant difference could be found between the two soil types in terms of texture, pH value, organic matter, and nitrogen, except for phosphorus and potassium level, which are higher in organic soil but higher overall in both soils according to the nationally defined limit. Chapter 6 assesses adoption of organic means of crop management practices, which in this chapter indicates soil and pest management practices, viz., mulching, compost shed, bio-slurry, vermicomposting, plastic cover, and bio-pesticides. These practices are evaluated against households' socioeconomic characteristics and analyzed the complementary or competing relation among themselves. Chapter 7 looks into crop diversification across these two farming systems using Shannon Diversity Index that captures both richness (number) and evenness (abundance) of crops and analyzes impact of households' socioeconomic characteristics on it. Organic farming in the study area is richer in integrating more number of crop types (richness) but is poor in evenness, which resulted in having lower index than conventional farming.

Chapters 8 and 9 relates with economic dimension of a farming system. Chapter 8 assesses farm income (total farm valuation) and gross farm cash income (income from selling crops in the market) generated among organic and conventional farmers and analyzes factors impacting them taking into consideration the existence of premium market. This chapter finds that income from organic farming is lower because production per hectare, commercialization rate, and price at which the crops are sold per unit are higher for conventional farming and access to premium market for organic products is also very limited. It also discusses why linkage with local organic market is necessary to improve income from organic farming. Chapter 9 assesses production and net return from three selected crops: carrot, potato, and cauliflower that are among the most commercial non-staple crops and form an important part of daily food consumption in the study area and throughout the country as well. Net return from conventional potato is significantly higher, but overall production of three crops combined is found to be higher in organic farming system. Among the production factors, land area under cultivation, seed, organic inputs, chemical inputs, and tillage have positive impact. Since chemical inputs can deplete soil fertility over time, it is emphasized that crop production should be improved by adopting a more sustainable way, such as boosting use of organic inputs.

This book also analyzes local organic market to have a general understanding of status of organic products which are mainly confined within urban areas (Chap. 10). A total of 15 categories of products were identified: cereals, vegetables, spices, pulses, oil seeds, fruits, tea, coffee, juice, wine, pickles, honey, jam, snacks, and skin care products. While most of the organic products are sold on the basis of word of mouth, some are certified by designated certifiers, while others are approved by various agencies as safe to consume. The premium is usually higher for certified organic products, but surprisingly some self-claimed organic products are found to have higher premium rate than those certified as well because of already established networks. This can be attributed to the fact that in Nepalese context, some farmers even without certification are able to get premium price purely based on mutual trust or personal links, whereas others are devoid of such benefit despite of being certified because of poor marketing system and skill.

Chapter 11 discusses result from field experimentation of organic production of three selected crops, viz., carrot, kidney bean, and potato, which were selected for their importance in the share of daily food consumption and commercial value as well. The study was conducted by associating with a farmers' group for organic farming. About 10 plots sized $7.5m^2$ each were allocated for each crop, among which two plots were allocated for experimenting crop production with and without irrigation, two plots for with and without mulching using straws, two plots for with and without using self-made bio-pesticides, and remaining four plots for pruning at the rate 0%, 25%, 50%, and 75%.

Lastly, Chap. 12 concludes the book by evaluating how organic farming fares in terms of these three dimensions of sustainability in the study area. It is found that there is no substantial difference between organic and conventional farming systems when measured in all three dimensions of sustainability: social, economic, and environmental. Such could be the situation in other developing countries as well where there is often a vague difference between organic and conventional farming systems in a sense that both incorporate integrated farming of crops and livestock. Socially, organic farming is able to form a social fabric through group formation where farmers can gain knowledge through training and interaction. Environmentally, the soil quality for both farming systems is poor in terms of texture, pH, micronutrients, phosphorus, and potassium level. Higher adoption rate of organic means of crop management practices in organic farming indicates it to be more sustainable overtime. The crop diversity as calculated using Shannon Diversity Index showed that organic farming scored lower than conventional farming. Economically, conventional farming tested better than organic farming as the former has higher farm income and earns higher farm cash income than the latter. One of the reasons to have higher farm income is because production per hectare is higher in conventional farming system. However, production for all three selected crops combined – carrot, potato, and cauliflower – was higher for organic farming. It indicated that while the aggregate crop production might be higher for conventional farming, it is not necessarily so on a crop-by-crop basis. Comparing average net return from the same three crops, potato and carrot gave higher net return in conventional farming, while cauliflower gave better net return in organic farming. The most established market for organic products within the country exists in Kathmandu and Lalitpur cities where premium ranges from 2% to 40%, but unfortunately farmers in Chitwan District have very little means to connect to such market.

Therefore, in this case, no one farming system is better in all aspects tested under three sustainability dimensions. However, the concept of time horizon has a role to play in understanding the concept of sustainability as usually environmental change takes place after many years of accumulated processes. Thus, agro-system that appears sustainable today could be in the process of being unsustainable over a long time period. While various studies have supported the negative consequences brought on by conventional farming and positive aspects of organic farming that takes a much more sustainable approach, we should be more careful while glorifying the benefits of conventional farming with only present context in mind.

With the understanding that organic farming is a more sustainable approach toward the future of food security, for a developing country like Nepal, designating organic production pocket areas especially by involving smallholder farmers in a group is essential for its growth and development. It not only serves as a basis of minimizing contamination of organic farm from restricted inputs, especially through conventional farms, but provides various other benefits. These benefits come because of the easier recognition of organic farming, which otherwise would have taken humongous resources for a single farmer to achieve. Group conversion gives platform to share information and ideas among farmers since organic farming is very much knowledge-intensive and an opportunity to take their grievances to, collaborate with, and get assistances from various stakeholders. It also allows opting for certifying at the lowest cost possible through tools like the Participatory Guarantee System. The government should play an equally important role by linking research institutes to work directly with farmers and to the potential market based on their expertise of crop production determined by their socioeconomic and ecological context. Especially those areas that are claimed as organic by default should be considered for immediate actions.

This book is expected to enhance understanding on status of organic farming in a developing country like Nepal and how it fares in three dimensions of sustainability compared to conventional farming at the given moment. Nevertheless, readers are highly welcome for any critical comments and suggestions that would assist in improving the future research. Lastly, we would like to express our deepest gratitude to many people and institutions for their generous contribution in completing this book. First of all, we are extremely thankful to Hiroshima University that provided us the platform in the first place to carry out this research. Special recognition goes to Global Explorers to Cross Borders (G.ecbo) internship program of Hiroshima University for the financial support and "The Fuji Xerox Setsutaro Kobayashi Memorial Fund" for the research grant, without which it would have been unimaginable to conduct the research of this extent. Thank you to the Forum for Rural Welfare and Agricultural Reform for Development (FORWARD) Nepal for facilitating the internship during which most of this study was carried out. Our gratitude also goes to Soil Testing and Service Section of the Department of Agriculture in Nepal for providing their professional service of soil testing.

We are sincerely grateful to Prof. Dharma Raj Dangol and Prof. Moha Dutta Sharma of the Institute of Agriculture and Animal Science (IAAS); Prof. Akinobu Kawai of the Open University of Japan; and Prof. Shinji Kaneko of Hiroshima University for their constructive suggestions during the analysis of the results. To the fellow members of Maharjan seminar of the Graduate School for International Development and Cooperation (IDEC), Hiroshima University, under the Division of Educational Development and Cultural and Regional Studies, we are very appreciative of your contribution in improving the findings of this book through your insightful discussions and suggestions. Special thanks to Mr. Chandra Prasad

Adhikari, an avid organic farmer, and his family for accompanying us during the survey and sharing knowledge on different issues pertaining to organic farming in the study areas. Last but not the least, to all the enumerators and respondents, we express our earnest gratitude for your generous participation. There are many others who have contributed in one way or the other to bring this book to its present form. We deeply appreciate your support and guidance throughout this endeavor.

Higashi-Hiroshima, Japan
Mrinila Singh
Keshav Lall Maharjan

Contents

1	**Sustainability of Farming System: An Overview**		1
	1.1	Introduction	1
	1.2	Sustainability of Farming System	3
	1.3	Scope of Organic Farming for Sustainability	5
	1.4	Organic Farming in Global Context	8
	1.5	Organic Farming in South Asia	9
		1.5.1 India	11
		1.5.2 Bangladesh	13
		1.5.3 Bhutan	14
		1.5.4 Sri Lanka	15
		1.5.5 Pakistan	16
	1.6	Summary	16
	References		17
2	**Status and Scope of Organic Farming in Nepal**		21
	2.1	Introduction	21
	2.2	Organic Farming in Nepalese Context	23
	2.3	Organic Farming in Response to Climate Change in Nepal	25
	2.4	Organic Farming in Response to Food Insecurity in Nepal	26
		2.4.1 Food Availability	27
		2.4.2 Food Accessibility	28
		2.4.3 Food Utilization	28
		2.4.4 Food Stability	29
	2.5	Institutional Role for Development of Organic Farming in Nepal	30
	2.6	Summary	33
	References		34
3	**Organic Farming in Chitwan District of Nepal**		37
	3.1	Introduction	37
	3.2	Source of Data	42

	3.3	Sample Design	43
	3.4	Households' Background	47
	3.5	Nature of Group Formation	51
	3.6	Summary	56
	References		57
4	**Socioeconomic Dimension of Farming System**		**59**
	4.1	Introduction	59
	4.2	Socioeconomic Variables' Relation to Adoption of Farming System	60
	4.3	Empirical Model	63
	4.4	Socioeconomic Factors Impacting Adoption of Farming System	64
	4.5	Summary	68
	References		68
5	**Soil Properties of Organic and Conventional Farming Systems**		**71**
	5.1	Introduction	71
	5.2	Soil Properties	72
	5.3	Methodology of Soil Testing	74
	5.4	Difference in Properties of Organic and Conventional Soil	75
	5.5	Summary	79
	References		80
6	**Crop Management Through Organic Means**		**83**
	6.1	Introduction	83
	6.2	Socioeconomic Variables' Relation to Adoption of Organic Means of Crop Management Practices	85
	6.3	Empirical Model	92
	6.4	Organic Means of Crop Management Practices	94
	6.5	Socioeconomic Impact on Adoption of Organic Means of Crop Management Practices	95
	6.6	Summary	99
	References		100
7	**Crop Diversification Under Organic and Conventional Farming Systems**		**103**
	7.1	Introduction	103
	7.2	Socioeconomic Variables' Relation to Crop Diversification	104
	7.3	Empirical Model	107
	7.4	Crop Diversity in Organic and Conventional Farming Systems and Influence of Socioeconomic Factors	108
	7.5	Summary	110
	References		111

8	**Income from Organic and Conventional Farming Systems**		113
	8.1	Introduction	113
	8.2	Insight on Farming-Related Income and Its Influencing Factors	115
	8.3	Empirical Model	119
	8.4	Socioeconomic and Farming System Impact on Farming-Related Income	121
		8.4.1 Farm Income	121
		8.4.2 Market Involvement	124
		8.4.3 Farm Cash Income	126
	8.5	Summary	129
	References		131
9	**Crop Production and Net Return from Organic and Conventional Farming Systems**		133
	9.1	Introduction	133
	9.2	Sample Selection and Data Collection	135
	9.3	Influencing Factors on Crop Production	135
	9.4	Empirical Model	140
	9.5	Production and Net Return of Selected Crops	143
		9.5.1 Production Factors, Production, and Net Return	143
		9.5.2 Factors Impacting Crop Production	146
	9.6	Summary	148
	References		148
10	**Status of Local Organic Market in Nepal**		151
	10.1	Introduction	151
	10.2	Data Collection	153
	10.3	Analysis of Local Organic Market in Nepal	156
	10.4	Summary	164
	References		165
11	**Field Experimentation of Vegetable Production**		167
	11.1	Introduction	167
	11.2	Field Setup for Experimentation	168
	11.3	Result from Field Experimentation	172
		11.3.1 Carrot	172
		11.3.2 Kidney Bean	173
		11.3.3 Potato	175
	11.4	Summary	176
	References		177
12	**Organic Farming from Perspective of Three Pillars of Sustainability**		179
Appendices			193
Index			199

Abbreviations

AAA	Appropriate Agricultural Alternatives
ACT	Organic Agriculture Certification Thailand
APEDA	Agricultural and Processed Food Products Export Development Authority
APP	Agriculture Perspective Plan
BADC	Bangladesh Agricultural Development Cooperation
BLM	Binary logistic model
BNOP	Bhutan National Organic Program
BOPMA	Bangladesh Organic Products Manufacturers Association
BPM	Bivariate probit model
BT	*Bacillus thuringiensis*
CAC	Codex Alimentarius Commission
CBS	Central Bureau of Statistics
CertAll	Certification Alliance
CWDS	Community Welfare and Development Society
DADO	District Agriculture Development Office
DAP	Diammonium phosphate
DFID	Department for International Development
DFTQC	Department of Food Technology and Quality Control
DoAE	Directorate of Agriculture Extension
EC	European Commission
EM	Effective Microorganisms
EOR	European Organic Regulations
EPA	Environmental Protection Agency
ESCAP	Economic and Social Commission for Asia and the Pacific
Euro GAP	Euro Good Agricultural Practice
FAO	Food and Agriculture Organization
FFS	Farmer's Field School
FiBL	Research Institute of Organic Agriculture
FYM	Farmyard manure

GDP	Gross domestic product
GHG	Greenhouse gas
GHK	Geweke-Hajivassiliou-Keane
GMOs	Genetically modified organisms
GPS	Global Positioning System
HDRA	Henry Doubleday Research Association
HH	Household
HHH	Head of Household
HORTEX	Horticulture Export Development
ICCOA	International Competence Centre for Organic Agriculture
ICS	Internal Control System
IFDC	International Fertilizer Development Center
IFOAM	International Federation of Organic Agriculture Movements
IFPRI	International Food Policy Research Institute
IIA	Independence of Irrelevant Alternative
INSAN	Institute for Sustainable Agriculture Nepal
IPCC	Intergovernmental Panel on Climate Change
IPM	Integrated pest management
IPNMS	Integrated Plant Nutrient Management System
ISFM	Integrated Soil Fertility Management
ISO	International Organization for Standardization
IUCN	International Union for Conservation of Nature
JAS	Japanese Agricultural Standard
JPP	Jajarkot Permaculture Program
LFU	Labor Force Unit
Ln	Natural log
LOAM	Lanka Organic Agriculture Movement
LSU	Livestock unit
masl	Meters above sea level
MNL	Multinomial logit
MoAD	Ministry of Agriculture Development
MoE	Ministry of Environment
MoHP	Ministry of Health and Population
MOP	Muriate of Potash
MVP	Multivariate probit
NARC	Nepal Agricultural Research Council
NASAA	National Association for Sustainable Agriculture, Australia
NCO	NASAA Certified Organic
NECOS	Nepal Community Support Program
NGO	Nongovernmental organization
NOAAB	National Organic Agriculture Accreditation Body
NOP	National Organic Program
NPC	National Planning Commission
NPG	Nepal Permaculture Group

NPOP	National Programme for Organic Production
NRs.	Nepalese rupees
NTCDB	National Tea and Coffee Development Board
OCMPs	Organic means of crop management practices
OCN	Organic Certification Nepal
OECD	Organisation for Economic Co-operation and Development
OLS	Ordinary least square
PGS	Participatory Guarantee System
SD	Standard deviation
SGS	Société Generale de Surveillance
SHDI	Shannon Diversity Index
SML	Simulated maximum likelihood
SOM	Soil organic matter
STATA	Data Analysis and Statistical Software
STT	Soil texture triangle
TPC	Third-party certification
UN	United Nations
UNCTAD	United National Conference on Trade and Development
UNEP	United Nations Environment Programme
US$	United States dollar
USDA	United States Department of Agriculture
VDC	Village Development Committee
VDD	Vegetable Development Directorate
VIF	Variation inflation factor
WFP	World Food Programme
WHO	World Health Organization
WTO	World Trade Organization

List of Figures

Fig. 1.1	Three pillars of sustainability	5
Fig. 1.2	Trend of organic agriculture	9
Fig. 1.3	Map of South Asia	10
Fig. 3.1	Map of Nepal showing ecological zones, regional boundaries, and study area	38
Fig. 3.2	Annual sales of chemical fertilizer in Nepal 1992/1993–2013/2014	40
Fig. 3.3	Map of Chitwan District showing study areas (VDCs)	42
Fig. 3.4	Organic and conventional farms without buffer zone in between	44
Fig. 3.5	Small-scale organic vegetable farming usually for home consumption only	45
Fig. 3.6	Pesticide used for pests in potato	45
Fig. 3.7	Farmer preparing for chemical fertilizer application	46
Fig. 3.8	Package of vegetable seeds received by member organic farmers who participated in the training	53
Fig. 3.9	Improved compost-shed with a partial financial assistance from an NGO	54
Fig. 3.10	Tricycle with a carrier provided by an NGO for collecting and selling organic crops	54
Fig. 4.1	Farmers inspecting cabbage during Farmer's Field School	66
Fig. 4.2	Farmers discussing facts and problems encountered from each plot during Farmer's Field School	67
Fig. 5.1	Soil texture triangle	75
Fig. 6.1	Improved compost-shed	86
Fig. 6.2	Farm yard manure that is kept in open, risking volatilization by sun or leaching by rainfall	86

Fig. 6.3	Self-made bio-pesticide prepared in a plastic bin using local resources	87
Fig. 6.4	Bio-pesticide available in the market	87
Fig. 6.5	Crop cultivation using plastic cover	88
Fig. 6.6	Organic means of crop management practices adopted by the respondents	95
Fig. 6.7	Organic means of crop management practices adopted across two farming systems	95
Fig. 7.1	Distribution of crop types under various categories and across two farming systems	109
Fig. 8.1	Visual difference in organic (left) and conventional (right) brinjal found in the local market	126
Fig. 8.2	Total crops produced (kg/ha) under two farming systems	128
Fig. 8.3	Commercialization rate of two farming systems	128
Fig. 8.4	Price per unit of crop under two farming systems	129
Fig. 9.1	Commercial cultivation of carrot in Phoolbari VDC	141
Fig. 9.2	Carrots being washed on a commercial scale	141
Fig. 10.1	Global market for organic food in 2012	152
Fig. 10.2	Local organic markets in Nepal	154
Fig. 10.3	Number of items under each organic product category	157
Fig. 10.4	Number of items packaged under each organic product category	157
Fig. 10.5	Organic claim of items under each product category	158
Fig. 10.6	Map of Nepal showing source of organic products and destination of sale	160
Fig. 11.1	Organic disease and pest management, Farmer's Field School	169
Fig. 11.2	Women farmers inspecting plot for carrot production	173
Fig. 11.3	Farmers gathered for studying plots for kidney bean	175
Fig. 11.4	Farmers weighing potatoes after harvesting from the plots	177

List of Tables

Table 1.1	Organic farming in South Asia	11
Table 3.1	Crops, livestock, and fishery productivity comparison in Chitwan District, central Tarai region, and Nepal	39
Table 3.2	Food availability and requirement in 2013/2014 in Chitwan District, central Tarai region, and Nepal	42
Table 3.3	Distribution of respondents belonging to two farming systems across VDCs and based on group membership	46
Table 3.4	Distribution of respondents across VDCs and based on group membership	46
Table 3.5	Descriptive analysis of (categorical) variables across two farming systems	48
Table 3.6	Descriptive analysis of (continuous) variables across two farming systems	49
Table 3.7	Farming practice after group formation	51
Table 4.1	Definition, measurement, and hypothesized relation of socioeconomic variables to adoption of farming system	62
Table 4.2	Result from binary logistic model and marginal effect for organic farming system	65
Table 4.3	Differentiating factors across village development committees	67
Table 5.1	Interpretation of soil parameters	76
Table 5.2	T-test of soil properties	77
Table 5.3	Estimates of phosphorus and potassium removal by vegetable crops	79
Table 6.1	Expected sign of socioeconomic variables against dependent variable of OCMPs	90
Table 6.2	Parameter estimates of multivariate probit model for organic means of crop management practices	96

Table 7.1	Expected relation of explanatory variables with respect to dependent variable of SHDI	105
Table 7.2	Result from ordinary least square model for Shannon Diversity Index	109
Table 8.1	Organic products sold by cooperative in Phoolbari VDC (April–May 2012 to March–April 2013)	115
Table 8.2	Definition and measurement of selected variables for farm and cash income	116
Table 8.3	Measurement and summary of dependent variables of market involvement and cash income	120
Table 8.4	Result from ordinary least square model for farm income	122
Table 8.5	Result from bivariate probit (selection) model for marketing crops	124
Table 8.6	Comparing farm cash income across two farming systems from six crops that were partly sold in the premium market	125
Table 8.7	Result from bivariate probit model for marketing crops and ordinary least square model for gross farm cash income	127
Table 9.1	Sample distribution across crops based on VDC and farming system	136
Table 9.2	Sample distribution across crops based on membership and farming system	137
Table 9.3	Definition and measurement of selected variables for crop production	138
Table 9.4	Hypothesized relation of independent variables to dependent variable of crop production	140
Table 9.5	Net-return calculation for selected crops under organic and conventional farming systems	144
Table 9.6	Organic carrots sold through cooperative in Phoolbari VDC (April–May 2012/March–April 2013)	146
Table 9.7	Result from ordinary least square model for production per hectare of crop production	147
Table 10.1	List of visited outlets	155
Table 10.2	Premium price based on organic claim of products	161
Table 11.1	Participants based on gender during field experimentation	169
Table 11.2	Crop plantation information	171
Table 11.3	Weekly temperature during crop growing phase	172
Table 11.4	Result from carrot production	173
Table 11.5	Result from kidney bean production	174
Table 11.6	Result from potato production	176

List of Appendices

Appendix I	Information on Formal/Informal Groups Formed for the Purpose of Organic Farming	193
Appendix II	List of Types of Crops Under Six Broad Categories Cultivated in the Study Areas	193
Appendix III	List of Organic Products Identified in the Local Market	197

Chapter 1
Sustainability of Farming System: An Overview

Abstract The term sustainability is gaining popularity worldwide across various entities, and farming remains no exception. Responsibility to feed the growing population has brought remarkable changes in the way we produce food. The green revolution, also known as conventional farming system, although is known to produce more, is criticized for its high energy and input use, disregard for the environment and health of living beings, and disparity of its benefits across the world. This has now questioned the sustainability of such practice. This chapter discusses about the social, economic, and environmental aspect of organic farming in the light of sustainability of a farming system. Based on literature review, it shows how organic farming can achieve these three dimensions of sustainability, which mostly has to do with the local context. The global scenario shows that organic farming is indeed in a growing trend. While South Asian countries' organic sector is mostly export oriented, its local market also seems to be on rise given increasing purchasing power and awareness of health impact of food residues among consumers.

1.1 Introduction

Farming sector has been continuously changing when it first began thousands of years ago. From hunting and gathering, people started domesticating plants and animals some 10,000 years ago which was then followed by sedentary farming. Traditional farming system did shifting cultivation where land was kept fallow in the cropping sequence for as long as 30 years to maintain soil fertility, remove weeds, and control plant diseases by allowing regeneration of a forest cover. With increase in population and requirement for more food production, farming slowly started to be intensified by first reducing the fallow area and then moving on to double cropping, intensifying use of inputs such as manure, labor, and irrigation, using varieties with shorter growing season, and so on (Wolman and Fournier 1987). With the onset of industrial revolution in the eighteenth century, it has gone through some massive transformation. It started in Great Britain during the early 1700s, reached North America by the mid-1800s, and continues to be transferred and evolve even today in many countries with varying pace. Animal draft and human labor started to be replaced by mechanized source of power; use of chemicals gave a new meaning to improved crop production, shifting from subsistence to mono-cropping; land consolidation maximized production efficiency; and now improved plant and livestock breeding through genetic engineering continues to revolutionize how farming is being practiced. What took previously centuries and generations to achieve certain goals could now be achieved within decades that significantly improved living standard of common people which earlier was only available to aristocracy (Saltveit 2003; IFPRI 2002). An example of English wheat that took 1000 years to increase from 0.5 to 2 tons/hectare (ha) but only 40 years to soar from two to six tons/ha is a common representative of such progress (Hazell and Wood 2008).

Our society has long been driven by the principles of economics where cost-benefit analysis has been the center of our decision-making, and it is through them that social and political issues in general are interpreted and actual policies and solutions assessed. But as a counterreaction, nonfinancial considerations such as concern for the natural environment and treatment of animals, respect for cultures and traditions, and cautiousness in technological innovations have forced us to rethink our rights and responsibilities toward the world (Loukola and Kyllönen 2005). The exploitative farming of combining more and more fertilizer with ever-higher-yielding varieties to increase the grain harvest is no longer performing as anticipated as well (Kesavan and Swaminathan 2008). As a result, the term "sustainability" is gaining enormous attention these days, not just in regard to natural environment but in far more complex entities such as human societies, cultural traditions, or social institutions. It has gripped in the area of public administration, politics, sciences, and society as a whole (Loukola and Kyllönen 2005). The concept of "sustainability" gained worldwide attention since the beginning of the twenty-first century. It started getting recognition with International Union for Conservation of Nature (IUCN) in 1969 and United Nations Conference on the Human Environment in Stockholm in 1972 where emphasis on possibility of economic prosperity without environmental degradation was much highlighted. Over the course, this concept has evolved through the World Conservation Strategy in 1980, the Brundtland Report in 1987, and the United Nations Conference on Environment and Development in Rio in 1992 (Adams 2006).

The term "sustainability" entails "sustenance, survival, or flourishing of a process, an organism, or a resource." However, this ideology is far from having a universally acceptable standard and is rather surrounded by conflicting principles as it includes a large variety of interacting factors in a complicated setting (Loukola and Kyllönen 2005). This multiple interpretation arises due to differences in short- and long-term time frames and challenges of trade-offs therein or one's own personal perception of it, which is again based on user's purpose, emotional intelligence, and values (Roth 2010). Among others, definition by Brundtland Report, which states sustainability as the one that "meets the needs of the present without compromising the ability of future generations to meet their own needs," remains the most widely accepted one. Though the path to achieve it is vague, it at least gives an acceptable meaning of sustainability in today's context, which is emphasizing on the need to work on reversing environmental degradation but at the same time not undermining the importance of economic development (Adams 2006).

1.2 Sustainability of Farming System

There is no doubt that the green revolution, which commonly is also known as conventional farming, profoundly increased crop production with the use of chemical fertilizers and pesticides and high-yielding varieties and contributed significantly in reducing world hunger. A good example can be of India where in the 1960s, Mendelian genetics and plant breeding accelerated the cereal production at a rate (2.8-fold increase) higher than the population growth rate (2.2-fold) during 1965–2000 (Kesavan and Swaminathan 2008). But overtime, we have started to realize that such production system has its own share of drawbacks in the form of environmental degradation, health implication, and imbalance of benefit entitlement across regions. Excessive and inappropriate use of chemical fertilizers and pesticides polluted groundwater, streams, rivers, and oceans, triggered land degradation through soil erosion, and severed deterioration of the arable soil. It caused professional hazard; killed beneficial insects and other wildlife; reduced biodiversity; increased pest adaptation and resistance, desertification, and water eutrophication; and affected those who consumed it through food residue (DFID 2004; Kassie and Zikhali 2009). Even in the case of India, since the mid-1980s, soil degradation mainly in irrigated agricultural regions along with yield stagnation have started appearing, while the population continue to rise. Conventional wheat-rice rotation system in regions of Punjab and Haryana has been largely responsible for the deterioration of soil quality and depletion of groundwater. Without the use of nitrogen-fixing crops in rice-wheat rotation, these regions that previously were known as granaries of India are slowly disintegrating into food-insecure regions (Kesavan and Swaminathan 2008).

Globally too, by the middle of the twentieth century, most industrialized countries had already achieved sustained food surpluses. Asia, which suffered hunger and mass starvation as late as the mid-1960s, also became self-sufficient in staple food within 20 years though its population more than doubled. However, Africa fell short of such success (Hazell and Wood 2008). When the world per capita agricultural production increased by 25% compared to 1960 level, Asia and Latin America were able to increase per capita food production by 76% and 28%, respectively, while it was 10% less per capita in case of Africa (DFID 2004). Growing starvation in developing countries and obesity and other health-related problems in developed countries make this farming system hugely flawed based on socioeconomic fairness. On top of this, because production is getting more specialized and concentrated within few large corporations, the environmental implication as a result of transferring food from the point of production to distribution is further putting pressure on the environment. Economically too, rising price of energy and agrochemicals have impacted economic profitability of such farming system

(Risku-Norja and Mikkola 2009). Therefore, increasing issues of food insecurity, resource degradation, and imbalance of benefits across regions have now raised the question of sustainability of conventional farming system (DFID 2004; Rigby and Caceres 1997).

Sustainability of a farming system means maintaining the quantity and quality of agricultural produce over a considerably longer period of time without an exhaustion of ecosystem services that crops and animals depend on for their productivity. Thus, farming can be sustained as long as ecological foundations such as soil, freshwater, biodiversity, renewable energy, and atmosphere remain intact. But anthropogenic activities have put exponential pressure on these systems, more so than the rate they can regenerate themselves. Excess population growth and usage of resources have led to growing damage to basic life support systems of land, water, forests, biodiversity, oceans, and the atmosphere (Kesavan and Swaminathan 2008). We now have come to the understanding that farming sector should not be evaluated based only on increasing crop yield and production. Use of energy, emissions of greenhouse gas, management of biodiversity, salinity of water and soil, sustainable use of water and maintaining its quality, soil health, interactions with river and groundwater systems, pesticide use, transgenic trait crop management, workplace health and safety, and consumer safety are some of the concerns that farming sector must address. On the other hand, climate change and variability, market forces, and change in government policy are external factors that it must tackle with simultaneously. Thus, farming sector is at the crossroad where it needs to increase food production to feed the growing population and supply quality food to meet the living standard of existing people that is only getting higher, while addressing the issues of sustainability (Roth 2010).

At present, sustainability is widely viewed from the perspective of social, economic, and environmental dimensions (Fig. 1.1). Even in case of farming, if it has degrading impact on one of these spheres, it cannot sustain over the long term (Roth 2010). Social sustainability refers to fairness to all people involved including farmers, consumers, and community at large. Economic sustainability is that farming activities should give monetary benefit that helps secure farmers' livelihood and others involved too by ensuring food security and being able to access other necessities of life such as education, health, and so on. Environmental sustainability means it keeps intact the environmental services provided through soil, water, and air that community relies on for their survival. These three dimensions intersect, impact, and influence each other either in a constructive or detrimental way. For instance, multiple cropping though provides social benefit through high-nutritional value crops and environmental sustainability through natural nutrient recycling processes, it may result in lower efficiency in economies of scale due to diversity of enterprises (Belicka 2005; FAO 2014) and thus might provide less economic incentive. Therefore, there needs to be a common ground where all of these dimensions can be achieved in a balanced way. For example, in the above case, better market mechanism for variety of crops rather than for selected few might help to certain extent in achieving economic sustainability as well.

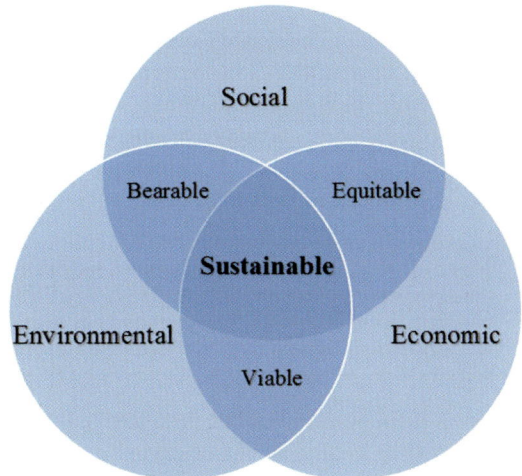

Fig. 1.1 Three pillars of sustainability (*Source*: Kates et al. 2005)

1.3 Scope of Organic Farming for Sustainability

The goal to sustainable farming is agreed upon universally, but it is not as easy to decide on one universal way that suits for all. Because there is such a range and number of parties involved, there will always arise a debate on deciding operational definition of sustainable farming more precisely. For instance, even agrochemicals can be claimed as contributing to farmers' economic sustainability (Rigby and Caceres 1997). An example of green revolution not being critical in its entirety, despite being criticized for numerous reasons, is that in some ways, it has actually saved forest area from being turned into a farmland. Green revolution of the 1960s and 1970s in India, for instance, led to substantial productivity increase which otherwise would have taken 80 million ha of more land in order to produce food grains at the present level (Kesavan and Swaminathan 2008). However, such benefit could not be realized ubiquitously. A study in sub-Saharan Africa identified lack of infrastructure, high transport cost, limited investment in irrigation, and unsuitable pricing and marketing policies to have hindered farmers to reap the benefits from such revolution (IFPRI 2002).Therefore, like the implication of green revolution in the above example, sustainable farming system is also both temporally and spatially bound, meaning sustainable practice claimed by one farmer at a given time may not be sustainable for another farmer at another point in time. Thus, an ability of a certain technology or system to be sustainable will highly depend on the distinctiveness of the context in which it is used (Rigby and Caceres 1997).

On the other hand, though green revolution is providing benefit in present context, long-term practice of such exploitative farming without complying with scientific principles has known to exhaust ecological foundations such as soil, freshwater, biodiversity, renewable energy, and atmosphere needed for sustainable farming productivity in quantity and quality over long period of time (Kesavan and

Swaminathan 2008). As a result, many alternative approaches such as integrated pest management, integrated crop management, low input agriculture, low input sustainable agriculture, low external input sustainable agriculture, agroecology, permaculture, biodynamic farming, and organic farming with respect to sustainability issues have been developed, but organic farming predates all other approaches for being an "environmentally friendly" farming (Rigby and Caceres 1997). According to IFOAM (2008):

> Organic agriculture is a production system that sustains the health of soils, ecosystems and people. It relies on ecological processes, biodiversity and cycles adapted to local conditions, rather than the use of inputs with adverse effects. Organic agriculture combines tradition, innovation and science to benefit the shared environment and promote fair relationships and a good quality of life for all involved. (Definition of Organic Agriculture, para. 3)

From this definition, it can be stated that organic farming (farming in this case is understood as a component of agriculture that is specifically related to producing crops and/or rearing animals) excludes use of chemical fertilizers and pesticides, genetically modified organisms, growth regulators, livestock feed additives, or hormones. Instead it relies on natural processes such as crop rotation, animal manure, green manure, natural enemies, pest-free plant varieties, companion planting, integrated pest management, and so on to control pests, weeds, and diseases and maintain health of soil and that of all living organisms involved as well. It emphasizes the use of locally available resources and optimum production under given environmental condition. Organic farming is known to be one of the most sustainable forms of production method. Many might perceive organic farming to be a traditional way of farming which used to be in practice before the introduction of green revolution. However, others understand it as incorporating best of traditional farming practices and assimilating them with modern scientific knowledge. Rather than relying completely on nature, organic farmers use all the knowledge, techniques, and materials available to work with nature. For example, combining green manure and careful cultivation is known to work best for weed control rather than using them separately (HDRA 1998). According to Vogt (2007), organic farming is "an intensification of farming by biological and ecological means in contrast to chemical intensification by mineral fertilizers and synthetic pesticides."

Benefit of organic farming has been reported widely on environmental, economic, or social grounds, either in isolation or collectively. For instance, SEKEM initiative, which started its first farm on biodynamic methods on 24 ha in the desert of Egypt 30 years ago, now has an organic certified area of almost 20 ha with another 20 ha in transition. Going organic has helped utilize rice straw which farmers usually burn, to make microbial compost, thus replacing mineral fertilizers. This has reduced emission of carbon dioxide, nitrous oxide, and methane gases, thereby increasing air quality. Economic liberalization policy and increased export to international market led farmers to rely on biological fertilizers instead of chemical fertilizer. A research showed organic farming significantly lowered nitrate-leaching rates per ha, lessening pollution in drinking water. Use of chemical pesticides reduced from 30,000 tons annually in the early 1990s to around 3000 tons

1.3 Scope of Organic Farming for Sustainability

in 2007. With better food quality, increasing awareness, and health consciousness, consumers' willingness to pay increased. Farmers too showed satisfaction due to significant reduction in health problems (Brandt 2007). A more comprehensive detail on results from numerous scientific studies has been compiled by Leu (2011) that validates organic method to be the low cost, high yielding, both environmentally and economically profitable endeavor. It has been claimed that even after introducing conventional farming, food production per person in Africa decreased by 10% compared to 1960s level. The United Nations Conference on Trade and Development (UNCTAD) and the United Nations Environment Programme (UNEP) found that organic farming can boost yields in Africa with crop yield increasing as much as 116% for all African projects. In Madhya Pradesh, India, farmers had to face declining returns, toxicity, and severe pest problem despite of increase in pesticide use, due to which many abandoned cotton production altogether. Then what started as an experimental plot for organic cotton farming, after 7 years, more than one thousand farmers were cultivating in more than 15,000 acres with cotton yields increasing up to 20% more than in neighboring conventional farms. Faced with similar problems in Peru too, organic cotton yielded 10–20% higher than the national average in arid coastal plain (Parrott and Marsden 2002). A similar result in China also showed improvement in food security in terms of nutrition and quantity, optimization of agricultural structure, and ensuring profit for both farmers and the company involved for organic vegetable production, processing, and trading (Brandt 2007). This proves that organic farming is not necessarily low yielding, though may be location specific. Thus, any judgment toward feasibility of organic farming in a given society and ways to overcome barriers for expansion of this sector should also be context specific.

Organic farming though known to be a sustainable approach to conventional farming, worldwide, its share as of 2014 was only 1% (43.66 million ha) of total agricultural land (Willer and Lernoud 2016). There are numerous reasons to why coverage of organic farming is significantly lower than conventional farming. Organic farming is most widely known for providing healthy lifestyle, either through residue-free food consumption or offering a hazard-free farming environment. It is known to adapt, mitigate, and being resilient to changing climate, which is a burgeoning issue in present-day scenario. It is also known to improve soil fertility for sustaining crop productivity overtime, be self-reliant by making use of locally available resources, reduce external shocks (climate or market related) through crop diversification, and allow farmers to enjoy higher profit through accessing premium price (Scialabba 2007). But on the contrary, it is claimed to produce less compared to its conventional counterpart, is labor-intensive, and, often times, is difficult to get hold of enough organically acceptable inputs as prescribed in guidelines of the said country (Trewavas 2002; Meisner 2007). Another reason is organic food is generally more expensive which makes it less desirable by those with low income.

There are number of explanations for its higher price. Firstly, organic food supply is less than demand. Higher labor cost per unit of output and lower efficiency in economies of scale due to diversity of enterprises also contributes to higher price.

Finally, no matter how limited the quantity of production is, it still needs to abide by the stricter regulation to maintain its integrity through process of certification and other postharvest requirements (processing, transportation, and marketing). Moreover, conventional food does not reflect the actual environmental and health cost incurred during its production and consumption phase (Belicka 2005; FAO 2014). Conventional farming though is known to cause environmental (polluting water source, land degradation, biodiversity loss), economic (increasing input cost, decreasing trend of output), and social concerns (occupational risk, food residue, regional disparities) (DFID 2004; Kassie and Zikhali 2009; Scialabba 2007), its ability to produce more (IFPRI 2002) makes it desirable as it meets the present demand. Then again, global food production in aggregate already surpasses requirement to feed the entire population at the lowest price we have ever known. Despite globalization and increasing world trade in agro-products, there is still the problem of spatial differences in societies' ability to feed themselves and protect long-term productive capacity of their natural resources (Hazell and Wood 2008). Thus, ability to feed through organic farming should also be discussed in terms of other support to transfer food from production to consumption point, in addition to its capability to produce consistently without degrading the quality of production resources.

1.4 Organic Farming in Global Context

According to survey by International Federation of Organic Agriculture Movements (IFOAM) and Research Institute of Organic Agriculture (FiBL), area under organic farming and market share has been increasing gradually throughout the world. By 2014, 172 countries were formally involved in this sector through complying with certain standard and managed by more than two million producers. An area under organic management reached 43.66 million ha or 1% of total agricultural land, an increase of about 39% in more than a decade. This can be attributed mainly to the increase in geographical coverage of data collection (172 countries in 2014 compared to just 120 countries in 2004). Actual depiction of its growth may be seen from its global sales, which reached US$80 billion in 2014, an increase of nearly 188% compared to more than a decade earlier (Willer and Yussefi 2006; Willer and Lernoud 2016). Thus, it can be implied that organic farming is an emerging sector and has been growing progressively (Fig. 1.2).

Oceania (39.7%) is the largest holder of organic agricultural land followed by Europe (26.6%), Latin America (15.5%), Asia (8.2%), North America (7.1%), and finally Africa (2.9%). Australia (17.2 million ha), Argentina (3.1 million ha), and the United States (2.2 million ha) have the biggest size of organic agricultural land in the world. Falkland Islands (Malvinas) (36.3%), Liechtenstein (30.9%), and Austria (19.4%) have the highest share of organic to total agricultural land. India (650,000), Uganda (190,552), and Mexico (169,703) have the highest number of producers. The United States (US$35.9 billion), Germany (US$10.5 billion), and

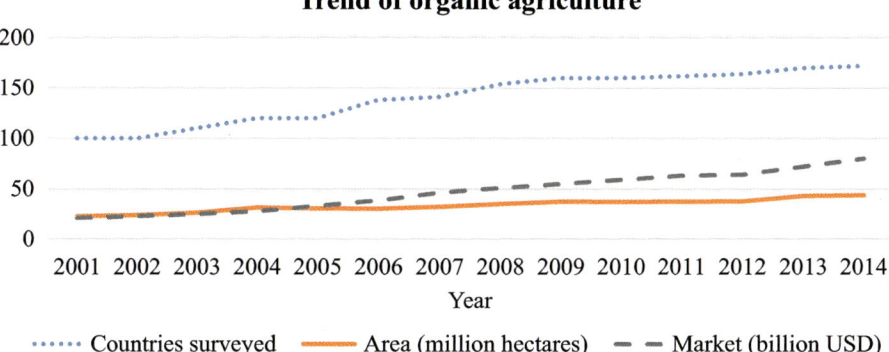

Fig. 1.2 Trend of organic agriculture (*Source*: Yussefi and Willer (2003); Willer and Yussefi (2004); Willer and Yussefi (2005); Willer and Yussefi (2006); Willer and Yussefi (2007); Willer et al. (2008); Willer and Kilcher (2009); Willer and Kilcher (2010); Willer and Kilcher (2011); Willer and Kilcher (2012); Willer et al. (2013); Willer and Lernoud (2014); Willer and Lernoud (2015); Willer and Lernoud (2016))

France (US$6.8 billion) have the world's biggest organic market size, while Switzerland (221 euros), Luxemburg (164 euros), and Denmark (162 euros) have the highest per capita consumption of organic products globally. In the case of Asia, in 2014, it comprised of 3.6 million ha of organic land, which is the growth of 4.7% from the previous year. It has 40% of the world's organic (0.9 million) producers and the third-largest market for organic products which is mainly export based. Within Asia, China has the largest area (1.9 million ha) under organic management, Timor-Leste has the highest proportion of organic agricultural land (6.8%), and India leads by the number of organic producers (650,000 producers) (Willer and Lernoud 2016).

1.5 Organic Farming in South Asia

Like any other region, South Asia also seems to be having its fair share of growth in an organic sector. South Asian region comprises of eight countries, viz., Afghanistan, Bangladesh, Bhutan, India, Maldives, Nepal, Pakistan, and Sri Lanka (Fig. 1.3). South Asia is one of the least developed regions in the world. With a population of 1.4 billion, it accounts for half of the world's poor. Countries like Afghanistan, Bangladesh, and Nepal fall under lower-income group; Bhutan, India, Pakistan and Sri Lanka under lower middle-income group; and only Maldives is considered to be in upper middle-income group (World Bank 2014). Agriculture is an important economic sector and still a large portion of its population is dependent on it for their livelihood in almost all the countries (accounting for 30% gross domestic product (GDP) and 80% labor force in Afghanistan (Kawasaki et al. 2012), 20% GDP and 52% labor force in Bangladesh (Thomas et al. 2013),

Fig. 1.3 Map of South Asia (*Source*: http://www.mapsofindia.com/maps-of-asia/saarc-country-map.html#)

16.8% GDP and 60% labor force in Bhutan (National Statistics Bureau 2012), 14% GDP and 60% labor force in India (BIOFACH 2014a), 3.99% GDP and 11.5% labor force in Maldives (Quandl 2014), 35% GDP and 65% labor force in Nepal (MoE 2011), 21.4% GDP and 45% labor force in Pakistan (Farooq 2013), and 13% GDP and 33% labor force in Sri Lanka (Chintana 2010)). South Asia accounts for 1.9% of the global organic share (829,499 ha) and 23.04% within Asia (Willer and Lernoud 2016). Table 1.1 shows current status of organic farming in South Asia (further explanation regarding each country is provided in Sects. 1.5.1, 1.5.2, 1.5.3, 1.5.4 and 1.5.5 below except for Nepal, which is discussed in more detail in Chap. 2).

1.5 Organic Farming in South Asia

Table 1.1 Organic farming in South Asia

South Asian countries	Organic (including in-conversion) agricultural land (in ha)	Share of organic agricultural land by country	Number of producers	Regulation on organic agriculture
Afghanistan	61 (2013)	0.0002% (2012)	264 (2012)	–
Bangladesh	6860	0.1%	9335 (2011)	In process of drafting
Bhutan	6829	1.4%	2680	Fully implemented
India	720,000	0.4%	650,000 (2013)	Fully implemented
Nepal	9361	0.2%	687 (2013)	In process of drafting
Pakistan	23,828	0.1%	108	In process of drafting
Sri Lanka	62,560	2.3%	524	–

Source: Willer and Lernoud (2016); Willer and Lernoud (2015); Willer and Lernoud (2014); Willer et al. (2013)
Note: Data is for the year 2014 unless otherwise indicated in parenthesis; Maldives is excluded because of insignificant role of organic farming in its economy in general

1.5.1 India

India has by far the most developed organic sector within South Asia. In 2014, it had 2nd highest increase of land under organic (0.2 million) worldwide, leading it to have 4th largest area of organic agricultural land in the world and 2nd highest organic agricultural area (0.7 million hectare) in Asia. Its organic area is 0.4% of its total agricultural area. It has the highest area under cereal (99.2 million hectares), citrus fruit (0.97 million hectares), dried pulses (28 million hectares), and vegetables production in the world, but there is no data on organic production. Likewise, it has among the largest temperate, tropical, and subtropical fruits but again lacks data when it comes to organic production. It has 3rd highest wild collection and beekeeping areas (3.99 million hectares) and 2nd largest area under organic oilseed production (130,000 ha) globally. About 97% of the global organic cotton fiber supply comes from just five countries, among which India supplies three-quarters (74.25%) of the total production, making it the largest supplier (Willer and Lernoud 2016).

It has the third highest number of organizations that are affiliated (44 affiliates) to IFOAM in the world. It is among those few countries whose government's accreditation procedures are accepted by the US Department of Agriculture for certification according to the US National Organic Program (NOP), even though they are not directly accredited by the US Department of Agriculture (Willer and Lernoud 2016), and European Union as well (Willer and Lernoud 2014). In fact, the government initiated accreditation program for certification bodies instituted by National Programme for Organic Production (NPOP) in India are duly recognized by most countries at par with other global organic products (Singh 2013). It is the only country in South Asia that has fully implemented standards that are officially endorsed as organic by IFOAM- Organics International, based on their equivalence with the Common Objectives and Requirements of Organic Standards. It is known to have the highest number of organic (650,000) producers in the world. It also is the leading country in the world in terms of producers (23,317) involved in Participatory Guarantee System (PGS), of which 21,240 producers are certified making it the highest in the world as well (Willer and Lernoud 2016). PGS is a locally driven quality assurance method based on trust, networks, active participation, and knowledge exchange (IFOAM 2014). It has the 3rd highest area (9442 ha) managed by PGS-certified producers (Willer and Lernoud 2016).

Though the Ministry of Commerce launched the National Organic Program in 2000 as a result of detrimental impacts of green revolution resulting in declining soil fertility and pest immunity, requiring higher amount of fertilizer and pesticide use, organic industry in India is almost entirely export oriented wherein majority of farmers are opting for organic farming because of the economic benefit rather than for sustainability as the main reason (Pandey and Singh 2012). In 2012, its retail sales reached 130 million euros, and in 2014, exports were worth 303 million euros (Willer and Lernoud 2016). According to Agricultural and Processed Food Products Export Development Authority (APEDA 2011), India's export of certified organic products grew by 33% from previous year to reach US$157.22 billion. India produces variety of organic products with 1000 certified products to its credit. It has the potential to supply all categories of organic products in an international market. Being one of the important players in global organic market, it has also been dubbed as the fastest growing organic food market in the world (Singh 2013). Domestic organic market is also growing rapidly at the rate of 30–40% annually. It is all attributed to rising disposable income and health consciousness among urban dwellers. India Organic Trade Fair, organized annually by the International Competence Centre for Organic Farming (ICCOA) has helped substantially for developing the local organic market. Including hypermarket, organic produces can also be seen in retail shops which devote separate shelve for organic products, producer-owned stores, informal *haats* (rural market), online stores, etc. Few labels such as Bio Suisse are also encouraging to minimize food miles by rejecting transportation via plane (Kilcher et al. 2008; Singh 2013) due to its relatively high level of carbon dioxide (CO_2) emissions (Gibbon 2009). The substantial increment in market demand for organic food comes from the megacities such as Mumbai, Delhi, Chennai, Bangalore, Gurgaon, and Pune where people are getting more affluent and simultaneously have become more health conscious (BIOFACH 2014a, b). There are 2000 known organic outlets in the domestic market. Further, to track and regulate the organic imports, implementation of TraceNet is expected to be expanded, which is a web-based certification tracking system in which each operator is assigned a bar code linked to global positioning system (GPS) indicator number that enables products to be traced back to the farming plot including certificates, operator sanctions, and appeals (Willer et al. 2013).

The government is actively promoting organic sector with financial assistance to organic farmers. Government's organic friendly policy has also induced private sectors to be involved through farming, certification, processing, retailing, exporting, and bringing in new innovation to sell (Singh 2013). The new central government of India is providing US$64 million to two organic development initiatives: state of Meghalaya plans to convert 200,000 ha to organic farming by 2020 and Sikkim is aiming for 100% organic production. It also announced US$16 million for organic market development in eight states of Northeast region and US$48 million for PGS for the period of 2015–2016 (Willer and Lernoud 2016).

Despite such level of development, organic sector is not free from perils. Area under organic cultivation decreased to 0.78 million ha in 2010 due to spread of BT cotton (bacterium *Bacillus thuringiensis*, a genetically modified variety of cotton) and nonavailability of non-BT seeds. Large farmers' groups were rejected for certification because of the contamination. In addition to that, unfavorable organic cotton price and termination of state-backed certification support schemes resulted in farmers discarding certification renewal. The mandatory web-based certification data management method or online traceability scheme called TraceNet applied since 2012 by APEDA also proved to be a hassle and discouragement for many farmers (Willer and Lernoud 2012).

1.5.2 Bangladesh

As of 2014, Bangladesh had 6860 ha organic area, comprising 0.1% of its total agricultural land. Cereals accounted for 101 ha (0.001% share), tropical and subtropical fruit for 10 ha (0.003%), and vegetables for 157 ha (0.03%) (Willer and Lernoud 2016). Though organic market is still in a growing stage with only limited outlets selling on limited scale, the number of producers and companies advertising their products as organic is ever increasing. However, authenticity of such claim is debatable as organic labeling is not regulated, Bangladesh Organic Products Manufacturers Association (BOPMA) came up with its own standard by establishing a certification body named Organic Bangladesh Limited. BOPMA is also actively involved in uplifting this sector through training farmers and producing fertilizer and pesticides and is on the verge of establishing number of outlets. Government provides incentive to private exporters for exporting goods, supports entrepreneurs, and boosts technology adoption and work toward food safety-related complications. Different organizations are carrying out research but without proper coordination. In Bangladesh, organic products are sold through various ways – special section in many conventional stores, contract farming, and direct sales from farmers at local market and urban outlets of Bangladesh Agricultural Development Cooperation (BADC), but it is still not able to cater fully to the domestic demand (Willer and Lernoud 2014).

Moreover, most of the organic shops are limited in the capital city, Dhaka. Excluding the special eco-friendly outlets run by nongovernmental organization

(NGO), there are very few shops where organic items are sold in a corner along with other traditional food items. It faces problem of lack of trust among consumers for uncertified self-claimed organic products (Sarker and Itohara 2008). Nevertheless, local demand for organic products is growing with increasing awareness among consumers. However, production and marketing method of organic products based on contract farming by private organizations, companies, or chain shops have reduced the price received by farmers (Hoque 2012). Rise in local demand is accredited to food safety and environmental concerns. Organization such as Horticulture Export Development (HORTEX) Foundation is also working toward providing service on production, postharvest management, certification, and marketing for local and export market (HORTEX n.d.). Bangladesh has a huge organic aquaculture land accounting for 28% of 33,800 ha area available globally. At present it exports tea, shrimp, and some herbal and medicinal products on a limited scale but has so much more potential. Major areas to be considered are production, supply chain, information on import and export, and developing national policy, standard, inspection, and certification method for organic products (Willer and Lernoud 2014). As of 2014, it is still in the process of drafting national regulation for organic farming (Willer and Lernoud 2016).

1.5.3 Bhutan

Bhutan is another unique example, which caused a worldwide stir by claiming that it intends to make the whole nation organic by 2020. With this intention, Bhutan National Organic Program (BNOP) is giving support through training, seeds, seedlings, and soil fertility improvement techniques (Willer et al. 2013). The Ministry of Agriculture and Forests endorsed a plan for the production and supply of bio-inputs including distribution facilitation through the government system to reach throughout the country (Willer and Lernoud 2016). Though support from government is high, human resources and technical facility are the limiting factors (Willer et al. 2013). Another challenge is farther distance to the market from rural areas, which discourages farmers to sell their produce (Katwal n.d.).

As of 2014, Bhutan had 6829 ha, comprising 1.3% (4th highest share in Asia) of its total agricultural area. Its wild collection area is 6315 ha, cereal is 1037 ha (1.9% organic share to total cereal production), temperate fruit is 0.1 ha (0.003%), and vegetable is 76 ha (0.7%) (Willer and Lernoud 2016). Currently, few products are exported and few organic retails shops are situated in the capital (Dorji 1999). Organic produce is perceived to generate better revenue for farmers and alleviate poverty. One of the major cash crops, red rice, is produced in high altitude and is claimed as a natural product (without certification). It is exported to the United States and Europe, especially Germany and the United Kingdom with around 100 metric tons (mt) annually with a good profit margin (Duba et al. 2008; Agrifood Consulting International 2007). However, demand is considered to be around 200 mt, and it has not been able to meet the target fully because of scattered nature of production, low yields, absence of certification method, and the need for documentation (Agrifood Consulting International 2007). Bio Bhutan, a private enterprise, has successfully opened niche markets in Asia, Europe, and the United States for organic certified lemongrass oil with premium price (Yangzom et al. 2008), which is Bhutan's only organic certified product (Agrifood Consulting International 2007). Bhutan has the highest share of organic to overall agricultural area (1.21%) within South Asia. Nevertheless, it is still a huge challenge to realize its vision of going 100% organic by 2020. Besides India, it is the only other country in South Asia that has fully implemented regulation on organic farming (Willer and Lernoud 2014). On December 5, 2015, it initiated Bhutan organic domestic assurance system and is reviewing its national organic standards to qualify for international recognition in collaboration with IFOAM – Organic International to qualify for international recognition (Willer and Lernoud 2016).

1.5.4 Sri Lanka

Sri Lanka has 62,560 ha of organic agricultural land, comprising 2.3% of its total agricultural area; which is the 2nd highest share in Asia (Willer and Lernoud 2016). It established government-competent organic labeling regulation (Willer and Kilcher 2010) and joined collaboration of private and government-linked certification bodies for low-cost inspection and certification under the label of Certification Alliance (CertAll) (Willer and Kilcher 2011). In Sri Lanka too, claims of organic to get advantage of growing market have been emerging. Organic tea, coffee, spices, and fruits are well developed than other commodities in this country. Certification for export and development of new markets have emerged in the light of food contamination issue but still awareness on organic remains low (Willer et al. 2013). It banned the use of glyphosate and subsidized organic fertilizer. Its Export Development Board promotes organic products by assisting exporters seeking new markets. Out of 78,502 ha under organic management, 62,560 ha is certified. Export is valued at US$228 million with 1346 mt in total. Products are exported mainly to the United States, Canada, EU, Japan, and Australia, while exports to Middle East are also growing. Domestic market is expanding from urban to rural communities. Major supermarkets offer organic products verified under PGS. Local certification, market assurance, and developing consumer confidence have all been a contributing factor in marketing organic foods. Third-party certification also exists and assurance through PGS also seems to be on rise. Lanka Organic Agriculture Movement (LOAM) is a national organic association involved in development of assurance system for domestic market (Willer and Lernoud 2016).

1.5.5 Pakistan

As of 2014, Pakistan has 23,828 ha, comprising 0.1% of its total agricultural area. Among this, cereals account for 10,271 ha (0.1% of total cereals grown), and tropical and subtropical fruits for 878 ha (0.2%) (Willer and Lernoud 2016). Currently, it exports fruits mainly to the Middle East, Sri Lanka, and Central Russian States, where quality standards are not as rigorous as in developed countries. But after the introduction of World Trade Organization (WTO), it faces threat of losing many international markets. Therefore, it has decided to introduce Euro Good Agricultural Practice (Euro GAP) for improving its farming standards applicable to international standards (Farooqi 2007). Unfortunately, specialized market for organic do not exist at this moment, which means that local organic farmers are not benefited through better returns (Mehmood et al. 2011). It is still in the process of drafting national regulation for organic farming (Willer and Lernoud 2016).

Overall, organic sector in Asian countries are known to be mainly export based. In case of South Asia, countries like India and Sri Lanka have highly export-oriented organic sector. Though there is dearth of information on local and export market for most of the countries, specifically Afghanistan and even Pakistan; overall trend of organic sector is seen to be on rise in national market as well due to increasing income and growing awareness of health benefits from consuming organic products (Willer and Lernoud 2014).

1.6 Summary

The industrial way of farming has brought massive improvement in crop production to satisfy the demand of growing population and improving living standards as well. However, measurement of farming performance is no longer confined to crop yield and production. Conventional farming system relies heavily on energy use, has polluted and degraded environment, threatened biodiversity, and created health hazards. In addition, there still exists imbalance of benefits brought upon by this farming system. All of these negative consequences have forced us to rethink what sustainable farming actually implies. Organic farming is recognized to be a sustainable alternative that relies on locally available resources rather than depending on harmful inputs. However, its ability to produce comparable to conventional farming system may be debatable, sustainability is widely accepted in terms of social, economic and environment dimensions. While both of these farming systems might not fully satisfy each of these dimensions on an equal basis, there has to be a middle ground which can meet these expectations in the best possible way in a given situation. Organic farming is highly regarded for social and environment sustainability, although its economic ability also sometimes can be favorable given proper support is in place. Thus, benefit of this farming system very much depended on the local context. Organic farming the world over has been growing gradually which can be seen in its rising market share. Within South Asia, while the organic sector is mainly export based, its demand is increasing locally as well, especially in urban areas where health consciousness and mounting purchasing power have contributed to this trend.

References

Adams W (2006) The future of sustainability: re-thinking environment and development in the twenty-first century. International Union for Conservation of nature (IUCN), Switzerland
Agrifood Consulting International (2007) Bhutan export market assessment and product identification study. Export intensification program for Bhutan SME sector project, Prepared for the International Finance Corporation- South Asia Enterprise Development Facility (SEDF) by Agrifood Consulting International. Maryland
APEDA (2011) Data for organic products 2010–2011. http://www.apeda.gov.in/apedawebsite/organic/PresentStatus.htm. Retrieved 10 July 2013
Belicka I (2005) Organic food: ongoing general aspects. Environmental friendly food production system: requirements for plant breeding and seed production (ENVIRFOOD), Latvia
BIOFACH (2014a) India organic: the market place for organic people. BIOFACH-India, Bangalore
BIOFACH (2014b) India: a strong growing organic market. BIOFACH INDIA, Bangalore
Brandt K (2007) Organic agriculture and food utilization. Newcastle University, United Kingdom
Chintana M (2010) Sri Lanka: the emerging wonder of Asia. Ministry of Finance and Planning, Colombo
DFID (2004) Agricultural sustainability. Department for International Development, United Kingdom
Dorji P (1999) Deciduous fruit production in Bhutan. Regional Office for Asia and the Pacific, Food and Agriculture Organization of the United Nations (FAO), Rome
Duba S, Ghimiray M, Gurung TR (2008) Promoting organic farming in Bhutan: a review of policy, implementation and constraints. Ministry of Agriculture, Council for RNR Research of Bhutan
Quandl (2014) Maldives: economy data. http://www.quandl.com/maldives/maldives-economy-data. Retrieved 8 July 2014
FAO (2014) Organic agriculture: FAQ. http://www.fao.org/organicag/oa-faq/oa-faq5/en/. Retrieved 5 July 2014
Farooqi A (2007) Potential of organic farming to alleviate poverty in Pakistan. Department of Environmental Science, Allama Iqbal Open University, Islamabad
Farooq O (2013) Pakistan economic survey 2012–13: agriculture. Ministry of Finance, Islamabad
Gibbon P (2009) European organic standard setting organisations and climate-change standards. OECD Global Forum on Trade, Paris
Hazell P, Wood S (2008) Drivers of change in global agriculture. Philos Transact Royal Soc B Biol Sci 363(1491):495–515
HDRA (1998) What is organic farming? Henry Doubleday Research Association, Warwickshire
Hoque MN (2012) Eco-friendly and organic farming in Bangladesh: international classification and local practice. Institut für Agrarsoziologie und Beratungswesen der Justus-Liebig-Universität Gießen, Giessen
HORTEX (n.d.) Hortex Foundation. http://www.hortex.org/index.htm. Retrieved 22 April 2014
IFOAM (2008) Definition of organic agriculture. http://www.ifoam.bio/en/organic-landmarks/definition-organic-agriculture. Retrieved 20 March 2014
IFOAM (2014) .Participatory Guarantee System (PGS). http://www.ifoam.org/en/value-chain/participatory-guarantee-systems-pgs. Retrieved 12 Jan 2015

IFPRI (2002) Green revolution: curse or blessing? International Food Policy Research Institute, Washington, DC
Kassie M, Zikhali P (2009) Sustainable land management and agricultural practices in Africa: bridging the gap between research and farmers. University of Gothenburg, Gothenburg
Kates RW, Parris TM, Leiserowitz AA (2005) What is sustainable development? Goals, indicators, values and practice. Environ Sci Policy Sustain Develop 47(3):8–12
Katwal TB (n.d.) Multiple cropping in Bhutanese agriculture: present status and opportunities. Regional consultative meeting on popularizing multiple cropping innovations as a means to raise productivity and farm income in SAARC countries, Kandy
Kawasaki S, Watanabe F, Suzuki S, Nishimaki R, Takahashi S (2012) Current situation and issues on agriculture of Afghanistan. J Arid Land Stud 22(1):345–348
Kesavan P, Swaminathan M (2008) Strategies and models for agricultural sustainability in developing Asian countries. Philos Trans R Soc Lond B Biol Sci 363(1492):877–891
Kilcher L, Eisenring T, Menon M (2008) Organic market development in Africa, Asia and Latin America: case studies and Summaries for national action plans. 16th IFOAM Organic World Congress, Modena
Leu A (2011) Scientific studies that validate high yield environmentally sustainable organic systems. Organic Fereration of Australia, Mossman
Loukola O, Kyllönen S (2005) Sustainable use of renewable natural resources: from principles to practices. Department of Forest Ecology, University of Helsinki, Finland
Mehmood Y, Anjum MB, Ahmad M (2011) Organic farming in Pakistan: organic agriculture and the environment. http://organicpk.blogspot.jp/2011/12/organic-agriculture-and-environment.html. Retrieved 14 July, 2011
Meisner C (2007) Why organic food can't feed the world?. http://www.cosmosmagazine.com/features/online/1601/why-organic-food-cant-feed-world. Retrieved 29 Oct 2014
MoE (2011) Status of climate change in Nepal. Ministry of Environment. Government of Nepal, Kathmandu
National Statistics Bureau (2012) Statistical yearbook of Bhutan. Royal Government of Bhutan, Thimpu
Pandey J, Singh A (2012) Opportunities and constraints in organic farming: an Indian perspective. J Sci Res 56:47–72
Parrott N, Marsden T (2002) The real green revolution: organic and agroecological farming in the south. Greenpeace Environmental Trust, United Kingdom
Rigby D, Caceres D (1997) The sustainability of agricultural systems. Institute for Development Policy and Management. University of Manchester, Manchester
Risku-Norja H, Mikkola M (2009) Systemic sustainability characteristics of organic farming: a review. Agronom Res 7(II):728–736
Roth G (2010) Economic, environmental and social sustainability indicators of the Australian cotton industry. University of New England, Armidale
Saltveit ME (2003) Agriculture since the industrial revolution. http://www.encyclopedia.com/topic/Agriculture_industry.aspx. Retrieved 20 Oct 2012
Sarker MA, Itohara Y (2008) Organic farming and poverty elimination: a suggested model for Bangladesh. J Organic Syst 3(1):68–79
Scialabba NEH (2007) Organic agriculture and food security. International Conference on Organic Agriculture and Food Security. Food and Agriculture Organization of the United Nations (FAO), Rome, pp 1–22
Singh B (2013) On the brink of an organic revolution. India Brand Equity Foundation (IBEF), Haryana
Thomas TS, Mainuddin K, Chiang C, Rahman A, Haque A, Islam N, Quasem S, Sun Y (2013) Agriculture and adaptation in Bangladesh: current and projected impacts of climate change. International Food Policy Research Institute (IFPRI), Washington, DC
Trewavas A (2002) Malthus foiled again and again. Nature 418:668–670

References

Vogt G (2007) The origins of organic farming. In: Lockeretz W (ed) Organic farming: an international history. CAB International, Oxfordshire, pp 9–29

Willer H, Kilcher L (2009) The world of organic agriculture: statistics and emerging trends. International Federation of Organic Agriculture Movements (IFOAM)/Research Institute of Organic Agriculture (FiBL), Bonn/Frick

Willer H, Kilcher L (2010) The world of organic agriculture: statistics and emerging trends. International Federation of Organic Agriculture Movements (IFOAM)/Research Institute of Organic Agriculture (FiBL), Bonn/Frick

Willer H, Kilcher L (2011) The world of organic agriculture: statistics and emerging trends. International Federation of Organic Agriculture Movements (IFOAM)/Research Institute of Organic Agriculture (FiBL), Bonn/Frick

Willer H, Kilcher L (2012) The world of organic agriculture: statistics and emerging trends. International Federation of Organic Agriculture Movements (IFOAM)/Research Institute of Organic Agriculture (FiBL), Bonn/Frick

Willer H, Lernoud J (2014) The world of organic agriculture: statistics and emerging trends. International Federation of Organic Agriculture Movements (IFOAM)/Research Institute of Organic Agriculture (FiBL), Bonn/Frick

Willer H, Lernoud J (2015) The world of organic agriculture: statistics and emerging trends. International Federation of Organic Agriculture Movements (IFOAM)/Research Institute of Organic Agriculture (FiBL), Bonn/Frick

Willer H, Lernoud J (2016) The world of organic agriculture: statistics and emerging trends. International Federation of Organic Agriculture Movements (IFOAM)/Research Institute of Organic Agriculture (FiBL), Bonn/Frick

Willer H, Yussefi M (2004) The world of organic agriculture: statistics and emerging trends. International Federation of Organic Agriculture Movements (IFOAM)/Research Institute of Organic Agriculture (FiBL), Bonn/Frick

Willer H, Yussefi M (2005) The world of organic agriculture: statistics and emerging trends. International Federation of Organic Agriculture Movements (IFOAM)/Research Institute of Organic Agriculture (FiBL), Bonn/Frick

Willer H, Yussefi M (2006) The world of organic agriculture: statistics and emerging trends. International Federation of Organic Agriculture Movements (IFOAM)/Research Institute of Organic Agriculture (FiBL), Bonn/Frick

Willer H, Yussefi M (2007) The world of organic agriculture: statistics and emerging trends. International Federation of Organic Agriculture Movements (IFOAM)/Research Institute of Organic Agriculture (FiBL), Bonn/Frick

Willer H, Yussefi-Menzler M, Sorensen N (2008) The world of organic agriculture: statistics and emerging trends. International Federation of Organic Agriculture Movements (IFOAM)/Research Institute of Organic Agriculture (FiBL), Bonn/Frick

Willer H, Lernoud J, Kilcher L (2013) The world of organic agriculture: statistics and emerging trends. International Federation of Organic Agriculture Movements (IFOAM)/Research Institute of Organic Agriculture (FiBL), Bonn/Frick

Wolman MG, Fournier FG (1987) The industrial revolution and land transformation. In: Grigg D (ed) Land transformation in agriculture. Wiley, Queensland, pp 79–109

World Bank (2014) South Asia: countries. Retrieved from http://web.worldbank.org/: http://web.worldbank.org/WBSITE/EXTERNAL/COUNTRIES/SOUTHASIAEXT/0,,menuPK:158850~pagePK:146748~piPK:146812~theSitePK:223547,00.html. Retrieved 5 July 2014

Yangzom K, Krug I, Tshomo K, Setboonsarng S (2008) Market-based certification and management of non-timber forest products in Bhutan: organic lemongrass oil, poverty reduction, and environmental sustainability. ADBI discussion paper 106. Asian Development Bank Institute, Tokyo

Yussefi M, Willer H (2003) The world of organic agriculture: statistics and emerging trends. International Federation of Organic Agriculture Movements (IFOAM)/Research Institute of Organic Agriculture (FiBL), Bonn/Frick

Chapter 2
Status and Scope of Organic Farming in Nepal

Abstract Agriculture being the source of food, income, and employment for majority of the population, Nepal government has always emphasized this sector for dealing with key issues of poverty alleviation and economic development. Its support however is mostly inclined toward conventional farming. Organic farming holds a lot of potential in Nepal due to exclusion of costly agrochemicals, ecological diversities, and higher labor availability in farming sector. Besides, it also tackles the issue of climate change, food insecurity, and negative impact of conventional farming that have started to be realized in areas where use of chemicals is higher. While government has started to realize its importance and has come up with consistent policies, lack of adequate and integrated research, extension, manpower, and other support on organic farming production, marketing, and input supply have hindered the development of organic farming.

2.1 Introduction

Nepal is a landlocked country in South Asia, bordered by China to the north and India to the south, east, and west. With an area of 147,181 square kilometers, it is the 93rd largest country in the world. It extends from 80°4′ E to 88°12′ E longitude and 26°22′ N to 30°27′ N latitude. Ecologically, it is divided into three zones: mountain, hill, and plain (Tarai) with an altitudinal variation ranging from 60 meters above sea level (masl) in the south to 8848 masl in the north within a distance of only 160 kilometers (kms) (MoE 2011; Bhattarai 2006). Although only about 20% of its total area is cultivable (33% is forested and rest of the part is mostly mountainous), agriculture sector is the backbone of Nepalese economy contributing 36% to the gross domestic product (GDP) and forming the source of income and employment for 66% of the population (SECARD 2011; MoAD 2015; MoE 2011).

Agriculture in Nepal is very diverse due to the range of agro-climate brought on by its vast variation in topography and altitude. For example, within hills there are many microregions with differing temperature, moisture, humidity, sunshine hours, rainfall, vegetation, and so on. A crop variety suitable for one microclimate may not be suitable for another. While this limits the transferability of agriculture technology in such varied microclimatic conditions, it also offers potentiality of growing range of products at any given time of the year. Off-season crops such as cauliflower, cabbage, broccoli, kale, turnip, potato, and tomato in Tarai become main season in the hill, for instance (Paudel 2016). Agriculture sector is basically characterized as being subsistence and integrated with 60% of the farmers categorized as smallholders (Karki 2015). These smallholder farmers have landholding of less than 0.5 ha, on average (Atreya 2015). According to Mallick (2012), 78% of farm households produce mainly for home consumption; about 21% produce almost half for self-consumption and another half for selling, while only 1% is producing mainly for selling. Overall, only 13% of output is traded in the market.

The production within this sector is dominated by cereal production (49.41%), followed by livestock (25.68%), vegetable (9.71%), forestry (8.1%), and fruit and spices (7.04%) (Karki 2015). Among the ten most commercialized (at the rate of more than 50%) products that aggregately contributes 77% to agricultural GDP are milk, rice, vegetables, buffalo meat, potato, wheat, goat meat, fruits, ginger and fish. While cardamom, tea, coffee, flower, cotton, cocoon, sugarcane, jute, mushroom, and tobacco are the ten most commercialized (at the rate of more than 90%) commodities that aggregately contribute 2% to agricultural GDP. The most exported products are lentil (29%) followed by cardamom (7%), wheat (6.7%), tea (6.5%), and vegetables and ginger (2.1%) (Mallick 2012). The mountain and hill region has limited agricultural land for cereal crop production specially rice, wheat, and maize but is known for animal husbandry (sheep, goat, and yak farming) for wool, meat, milk, and milk products. Tarai, on the other hand, covers 15% plain area of the country. Having a subtropical climate, it has high potentiality of commercial farming production and is known as "the food bowl" of the country. In spite of that, even this region suffers from food deficiency as was reported for the year 2013/2014 (Paudel 2016).

Although from 2005/2006 to 2014/2015, average agriculture growth rate of 2.9% surpassed the population growth rate of 1.35% per annum, there has been a huge variation in growth from 1 year to the next. For example, in 2012/2013, the growth rate declined drastically to 1.1% from 4.6% in 2011/2012. This extreme variation has mainly got to do with it being highly weather dependent (Karki 2015; Paudel 2016). Thus, agriculture though is one of the major economic sectors of Nepal and despite of directly engaging more labor force than any other sector, it is highly underdeveloped. Compared to experimental plots, the farm level productivity of most food crops and livestock fall below 50% of the attainable potential. For example, the average attainable yield of major staple crops is 4–6 ton/ha, whereas the national average is only around 2.5–3.0 ton/ha. Besides being highly weather dependent, other drawbacks of this sector are low investment leading to sluggish development and transfer of technology, subsistence and scattered farming, poor linkage with research and extension, limited subsidies and support in agricultural inputs, rapid depletion of natural resources (declining soil fertility, loss of land due to land sliding, erosion and deforestation), labor scarcity, and limited trained human resources (Paudel 2016).

Agriculture being the source of food, income, and employment for majority of the population, Nepal government has always emphasized this sector for dealing with key issues of poverty alleviation and economic development. Among others, commercialization of farming has been deemed necessary to bring much needed changes in the economic growth of Nepal (Samriddhi 2011). Promoting farming commercialization and modernization for increased efficiency and farm income has been one of its major agriculture priorities (Karki 2015). Despite efforts to uplift this sector, agricultural productivity growth rate remains lower compared to other countries (Samriddhi 2011; Ghimire et al. 2012). In 2012, rice, the most important staple crop in South Asia, had the lowest yield in both Nepal and India (both around 1.8 mt/ha) compared to Bangladesh (approx. 2.9 mt/ha), Sri Lanka (approx. 2.7 mt/ha), and Pakistan (approx. 2.6 mt/ha) (Bishwajit et al. 2013). Such input-intensive farming system is known to degrade environmental services and stagnate or decline production overtime due to intensive and mono-cropping pattern method (Samie et al. 2010). Problems related to this kind of farming system is now emerging through declining soil fertility and production in those areas of Nepal which have history of long-term use of chemical fertilizers and pesticides (Bhatta and Doppler 2010; Weiss 2004; Shrestha and Neupane 2002).

Declining soil fertility, negative repercussions on environment and health of farmers due to use of agrochemicals and market demand reinforced organic movement in Nepal (Bhatta and Doppler 2010; Weiss 2004), and it has been growing gradually (Adhikari 2011). Organic farming is conceived to be a sustainable approach to food production method, an alternative to ecologically unsound practices of conventional farming. The already evident effect of climate change through declining food productivity (WFP 2009) calls for a more resilient agricultural practice such as organic farming to withstand unpredictable climate and assure production sustainability which is likely to be economically and socially just as well.

2.2 Organic Farming in Nepalese Context

Agrochemical application was initiated late as fertilizer was not known in Nepal until the early 1950s (Shrestha 2010). Before that, Nepalese farmers were practicing traditional farming that relies on local resources for crop production by integrating crops and livestock. A farming family would usually have a few pair of goats and chickens, a pair of bullock, few cows or buffalos, and others. Organic manure from these livestock were incorporated to the crop field, and crop residues were either incorporated into soil or fed to livestock (Atreya 2015). In order to intensify crop production, systematic channel for importing and distributing fertilizer began with the setup of Agriculture Input Corporation in 1966. Its usage significantly increased by 44.24% (from 108,730 mt in 1997/98 to 156,830 mt in 1998/1999) the year after the implementation of Fertilizer Deregulation Policy 1997 by involving private sector in fertilizer trade. However, over the years, the supply from formal source declined right after 1999 due to the rise in price of fertilizer in an international market and uncontrolled flow of cheap Indian fertilizers. It was estimated that fertilizer from such informal source accounted for two-thirds of the total supply volume. Thus, the overall supply is still claimed to have increased throughout these years. In Nepal, unbalanced use of fertilizer in terms of excessive use of nitrogenous fertilizers and unbalanced nutrient supply is rampant in areas that have easy access, which has increased soil acidity and deteriorated soil physical condition and underground water quality. This is especially true for Tarai where fertilizer use is four times higher than that in the hills because of easy access, close proximity to India and lower fertilizer prices (Shrestha 2010). Besides that, it has also resulted in environmental pollution; increased pest resistance; revived new plant pests; degraded forest area; intensified flooding, erosion, and drought; and declined overall productivity of major food commodities (Bhatta 2010; KC 2006). Now the importance of sustainability in farming has been realized and as a part of which organic farming started to get gradual recognition.

As of 2014, the total share of organic to overall agricultural land including in-conversion areas is reported to be only 0.2% (9361 ha) in Nepal which is an increase from 0.001% (45 ha) in 2001. The available data shows that it had 1183 farms in 2004, 687 producers in 2014, and 4 processors and exporters in 2010. However, this is a highly underreported data as much of organic farming has not been formalized yet (Yussefi and Willer 2003; Willer and Yussefi 2006; Willer and Kilcher 2012; Willer and Lernoud 2016). In high mountain and middle mountain areas, farmers still rely on traditional knowledge and locally available resources and are largely claimed as "organic by default" (Pokhrel and Pant 2009). "Organic by default" is the term used for farmers practicing organic farming compulsorily either because of geographical isolation or due to financial inability to purchase inorganic inputs. This so-called organic by default area is estimated to be around 800,000 ha (26%) of total agricultural land, which means that they are free from synthetic fertilizers and pesticides (Atreya 2015). Thus, it would not take much of an effort to convert their farming into organic, and by introducing organic farming principles, production can further be boosted by enhancing soil structure and fertility through nitrogen-fixing legume crops, improving composting techniques, and practicing self-made bio-fertilizer and bio-pesticides. Besides that, ecological richness has given Nepal with another advantage of producing quality organic fruits, vegetables, tea, coffee, cardamom, vegetable seeds, mushroom, honey, and medicinal plants and herbs. Some organic products like tea, coffee, honey, large cardamom, ginger, and medicinal herbs are already exported as well (Pokhrel and Pant 2009; DoAE 2006; Tamang et al. 2011; Paudel 2016). Among other products, Nepal is known to have the second highest (46%) share of organic to overall coffee production within the country (Willer and Lernoud 2016). The key export market for Nepal is Japan, South Korea, and India (Willer and Kilcher 2010).

After Nepal became a member of WTO, it has further increased prospect in an international market as "organic produce" is identified as potential export crop (Bhandari 2006; Pant 2006). Domestic market is also on rise with some existing even in mountain areas and price can range from 10% to 200% more than conventional products depending on the market location, quality, and the product itself (Willer and Kilcher 2009). Diversity of market channels such as ad hoc organic bazaars, small retail outlets, supermarket corners, multilevel direct selling, and internet marketing are thriving even from rural Nepalese markets (Willer and Kilcher 2010). Overall, organic farming is claimed to have huge potential in Nepalese context due to exclusion of costly agrochemicals, ecological diversities, and higher labor availability in farming sector (Pokhrel and Pant 2009), thus making it self-reliant.

2.3 Organic Farming in Response to Climate Change in Nepal

Climate change is a global issue, and it is indeed changing at a faster pace. According to Intergovernmental Panel on Climate Change, since 1850 when instrumental record of global surface temperature began, 11 of the last 12 years (1995–2006) were among the warmest years. Impact of climate change is clearly evident from increases in global average air and ocean temperatures, widespread melting of snow and ice, and rising global average sea level. The primary cause of such rapid change has been attributed mainly to anthropogenic activities. CO_2 concentration is a result of fossil fuel use and land use change, while methane (CH_4) and nitrous oxide (N_2O) are primarily due to farming practice. In the year 1994/1995, Nepal was responsible for 9747 greenhouse gas (GHG), 948 GHG, and 31 GHG of CO_2, CH_4, and N_2O gases, respectively. The major sources of emission were from enteric fermentation (28% of the total CO_2 equivalent) and agricultural soils (21.3%) followed by land use change and forestry sector, rice cultivation, biomass burnt for energy, manure management, and municipal solid waste disposal. When it comes to Nepal's share of per capita CO_2 equivalent emission, it remains below two metric ton, which is almost half as compared to the global average value of 3.9 mt per capita. The annual compound growth rate of CO_2 equivalent emissions from Nepal is 2% per annum, which is lower than emission by most developing countries. Nepal has always been vulnerable to climate variability, and global warming has further exacerbated its level of exposure. In a recent global risk index of the most vulnerable countries in the world, it is ranked fourth in the context of climate change and is mainly contributed to its more than 80% area being fragile ecosystem and having low per capita income. Adverse impact of climate change is already evident in water resources and farming and is likely to affect land use and land cover, biodiversity, ecosystem, human health, and livelihood method. Increasing intensity and frequency of flood, hailstones, landslides, soil erosion, drought, crop diseases, and increased temperature and varying precipitation pattern due to climate change have exacerbated the production loss (MoE 2011).

Agriculture sector thus being one of the major sources and taker of climate change, a more sustainable farming practice should be adopted which can blend with both of these aspects. Organic farming is considered to be the most sustainable food production system that has huge potential for mitigation and adaptation to climate change (Niggli et al. 2007). The main reason for less emission from this farming system is due to its avoidance of energy intensive inputs such as synthetic fertilizers. It rather uses leguminous crops, crop residues, and cover crops to supply nitrogen and improve soil fertility. Besides, improvement in soil fertility helps in higher and stable soil organic matter. Usually under organic farming, carbon dioxide sequestration into the soil is high which means less is being emitted into the atmosphere. The high rate of carbon retained in soil also has higher capacity to resist challenges to climate change. The soil's capacity to retain more water in organic farming makes it more productive during times of drought or in areas with erratic climatic conditions such as higher temperature and uncertain rainfall. Organic farming can complement well the agroforestry production system by increasing its productivity and at the same time reducing GHGs emission. Due to huge diversity, soil fertility, and most importantly relying on traditional skills and farmer's knowledge, organic farming is well equipped for adapting to the changing climate (Niggli et al. 2007). The vulnerability of soil to erosion, which is also one of the sources of GHGs emission, is also considerably reduced due to high level of organic matter and increase of carbon stocks. Preference for crop rotations and for seeds and breeds with better resistance to pest, diseases and climatic stress; lower fluctuation in yields; and diversification help reduce production risks in cases of a single crop failure, environmental adversity or socioeconomic shocks (Scialabba 2007). Thus, organic farming is a climate-friendly farming practice as it emits much lower level of GHGs and is known to quickly, affordably, and effectively sequester carbon in the soil. In addition to mitigating climate change, its other features like water efficiency, resilience to extreme weather events, and lower risk of complete crop failure make it one of the effective ways to adapt to changing climate as well. Given that soil is the major storage of carbon and that it accounted for a tenth of anthropogenic CO_2 emission since 1850, a change in agricultural practice to organic farming can help inverse this trend (IFOAM 2009).

2.4 Organic Farming in Response to Food Insecurity in Nepal

Environmental degradation and food insecurity are the major problems Nepal is facing today. Nepal has population of more than 26 million with annual growth rate of 1.4% (CBS 2011) and is considered to be one of the poorest countries in the world. Though it has achieved remarkable progress over the last few years, about 25% of its people are still living on less than US$1.25 a day (World Bank 2015). Food insecurity largely looms throughout the country. Around 25% of the households are considered to have insufficient food consumption to ensure a basic diet. Nearly half of the children under 5 years of age suffer from chronic undernutrition, while 15% of children under 5 years of age is affected by acute undernutrition (CBS

2013). Therefore, responsibility to feed its ever-increasing population remains highly significant (Shrestha 2010). In addition to it, problem of environmental degradation is getting apparent in farming sector. Studies show that land area covered with acidic soil rose from 36% to 61%; almost 98% of the soil is deficient in organic matter; and the amount of available nitrogen (N), phosphorus (P), and potassium (K) in soil is also decreasing. Additionally, more than 10,000 ha of agricultural land in Nepal have already been deserted, and although no specific data is available, hardening of soil, drying of water sources, and erosion of agro-biodiversity are frequently experienced by the farmers (Bhatta 2010).

Food security, as defined by FAO, is the physical and economic access to sufficient, safe, and nutritious food which meets individual's dietary needs and food preferences for an active and healthy life (FAO 2006). This means that food security not only implies enough food production but also accessibility and not only quantity consumed but the quality as well, for one to remain active and healthy. Food availability, accessibility, utilization, and stability are the most widely accepted dimensions of food security. While contemplating about capability of organic farming to feed the ever-growing population, we should look into how it can satisfy all these dimensions.

2.4.1 Food Availability

Food availability is "the availability of sufficient quantities of food of appropriate quality, supplied through domestic production or imports (including food aid)." This entails that there must be adequate quantity and suitable quality of food available at any given time. In case of organic farming, it is claimed that during the conversion phase of the first 2–3 years, yield is usually lower or sometimes indifferent if a low-input system has been converted. After the conversion period, organic yields have a potential to be more productive than conventional system. However, yield reduction is higher if the land had been run on a high-input level, and after conversion too, the productivity is generally lower than the conventional yield (Zundel and Kilcher 2007). In Nepalese context, it is further argued that chemical fertilizer usage is comparatively lower than in other countries (Vaidya et al. 2008) and thus comparatively will have less undesirable effect on production when changed to organic as loss is higher in areas where chemical has been used intensively before (Zundel and Kilcher 2007). Also places such as Kathmandu Valley and some districts in Tarai, where regular chemical fertilizers are applied, have started to experience deteriorating soil quality (Vaidya et al. 2008). Soil quality is directly related to soil health which largely determines productivity and environment health (Kinyangi 2007) that can be revived by organic farming. On the other hand, there are also places in Nepal where resource poor and subsistence farmers have been practicing farming in a traditional way which can be claimed as organic by default and thus converting to organic would not be that complex (Pokhrel and Pant 2009). This can be of particular interest for organic certification for export orientation. Besides these, there might be other indicators which decide the productivity under organic farming such as availability of water and required nitrogen-fixing materials for a complete soil fertility management system, labor cost, and assurance of receiving premium price, market accessibility, long-term contracts, and capacity building for farmers to benefit from the conversion. Clearly in Nepal, a lot needs to be done to fulfill the void in all of these facets.

2.4.2 Food Accessibility

Food accessibility is an "access by individuals to adequate resources (entitlements) for acquiring appropriate foods for a nutritious diet." Access to food depends on affordability of production inputs, livelihood factors such as income, and accessibility to food even by those far from the market area. Productivity does not assure accessibility. However, organic farming through household farm yields, nutritional benefits, cost-effectiveness, environmental sustainability, water and energy efficiency, biodiversity, labor availability, resilience, farmers' capacity, and social stability ensures accessibility to an extent possible. The probability of higher income and reduced debt, most likely due to exclusion of acquiring expensive fertilizers, poses financial benefits as well (Sligh and Christman 2007). Nepal government has always emphasized on chemical fertilizer to boost agricultural productivity. Areas where unbalanced (excessive) use of fertilizer is rampant have known to contribute in increasing soil acidity and deteriorating soil physical condition and underground water quality (Shrestha 2010). Since all of these fertilizers are imported, the major advantage of going organic would be to be dependent on locally available resource inputs combined with local knowledge, thus building farmers' resilience. The fact that over 80% of the population is involved in agriculture sector for livelihood (Rahija et al. 2011) makes it suitable for organic farming as it is known to demand more labor than conventional farming due to nonuse of chemical fertilizers and pesticides or drugs that has instant reaction on plant growth. A higher return on organic farming produce also enables food security through higher purchasing power for which there needs to be a strong market mechanism that ensures better price and accessibility as well. Besides these, the enhanced environmental services and energy efficiency in times of crisis makes organic farming a reliable option.

2.4.3 Food Utilization

Food utilization is the "utilization of food through adequate diet, clean water, sanitation and health care to reach a state of nutritional well-being where all physiological needs are met." Chemically induced farming results in professional hazard, environment problems, yield loss due to pests' resurrection, and food residues which has grave repercussion on farmers' and consumers' well-being. Though there is no comprehensive report of health hazards due to chemical exposure or consuming contaminated food, some studies have emphasized on this issue through haphazard use of chemical pesticides, food residue, and probable health issues due to such exposure or contact (Palikhe 2002; Shah and Devkota 2009). There has also been an incident when the lack of regular pesticide monitoring at the field level in Nepal resulted in harming export of tea, honey, and other food commodities in recent years (Koirala et al. 2009). Organic system contributes in proper food utilization by protecting from exposure to pesticides and providing safe, residue-free, high-quality nutritious food. The cropping diversity in organic farming along with rotation crops of minor economic value but that has high micronutrient and protein content such as legumes (Scialabba 2007) enhances household diets and hence improves health status. The higher activity of plant defense mechanisms and longer shelf life reduces losses during transport and storage (Brandt 2007). Thus, it greatly contributes to human health and that of the local environment.

2.4.4 Food Stability

Food stability is "both the availability and access dimensions of food security." It largely depends on natural resources and environmental services, climate change, and economic factors like trade policies. Climate change is now widely recognized as the major challenge facing the world today and agriculture sector is one of the major contributors to climate change. It accounted for 10–12% of total global anthropogenic GHGs emissions (Smith et al. 2007). On the other hand, farming certainly is extremely vulnerable to climate variability and change as it is highly dependent upon weather and climatic conditions to produce food and fiber required to sustain human life. Climate change is likely to have varying effect in the production rate of different regions. For instance, it is expected to favor temperate regions but hamper tropical ones (FAO 2008). Though higher yields in temperate regions might offset lower yields in tropical regions to maintain the global food productivity, it might not ensure food security at the local level, especially in developing countries where poor subsistence farmers might not have access to sufficient amount of food from other parts of the world where production might be increasing. The effects of climate change are already evident through reducing yields, particularly in developing countries due to rising temperature and decreasing water availability (Niggli et al. 2007), which calls for a resilient farming system to cope with changing climate. Agriculture sector should both be able to adapt to changes and provide a foundation for mitigating GHGs emissions as well as increase productivity to meet the global needs. Organic farming is inclined more toward this goal which is deemed to work with nature rather than against it. It is the most sustainable form of production system that is resilient and can adapt and mitigate climate change (IFOAM 2009). Its GHGs emission rate is much lower, and carbon sequestration in the soil is much higher. It builds resilience mainly due to efficient use of water, flexibility to withstand extreme weather events and minimize possibility of complete crop failure. Thus, due to resilience, soil stability, water-use efficiency, diversity, mitigation, and adaptation features have made organic farming competent to provide food stability for local community.

In Nepal too, food security is very much affected by natural adversities. The drought situation in the year 2005 and 2006 decreased food production (Regmi 2007) and so did during the time of 2008/2009 global food crisis. In addition to soaring food prices by 20–40%, Nepal was also hit by winter drought, severely affecting the household production. Likewise, flooding events in Tarai are also reported to be on rise and more vicious than before. This in addition to frequent crop loss due to disease and insect infestation has been causing agricultural losses (WFP 2009). Thus, for climatic stress-stricken country like Nepal, organic farming provides groundwork for increasing the resilience in agroecosystems to intensifying weather extremes such as drought, irregular rainfall, floods, and rising temperature. Also being a net importer of food (Paudel 2016), a self-reliant farming system which authenticates sustainability must be adopted for continuity in food supply. Techniques such as crop rotation and mixed cropping manage biodiversity, diversify and optimize farm productivity, reduce the need for procuring inputs, and stimulate market orientation among farmers for higher income. It contributes in soil stability and resilience of food supply through agroecological approach, which is an important factor in food supply stability. Organic farming system invests in natural and human capital – improving livelihood, providing fair return on labor, and offering affordable technology for boosting productivity (Scialabba 2007). Nevertheless, the proper extension services to bring out these numerous benefits should not be underemphasized.

2.5 Institutional Role for Development of Organic Farming in Nepal

The history of how organic farming was institutionalized in Nepal has been comprehensively analyzed by Atreya (2015). The formal initiation of organic farming in Nepal is relatively shorter. It was an American research scholar Miss Judith Chase who came to Nepal and started a small "organic garden" in Gamcha village of Bhaktapur district in 1987. She soon realized the potential of commercial organic farming and founded Appropriate Agricultural Alternatives (AAA), an NGO devoted to the promotion, research, advocacy, and marketing of organic farming in Nepal. Another organization named Institute for Sustainable Agriculture Nepal (INSAN) that was established in 1986 has been equally influential as well. It focused on the philosophy of permaculture and ecological farming, which are highly related to the principles of organic farming. During early 1990s, other NGOs such as Nepal Community Support Program (NECOS – estd. 1989), The

2.5 Institutional Role for Development of Organic Farming in Nepal

Lotus Land Agriculture Farm (estd. 1991), Jajarkot Permaculture Program (JPP – estd. 1991), Community Welfare and Development Society (CWDS – estd. 1992), Hasera Agriculture Research and Training Center (estd. 1992), and Ecological Service Centers (estd. 1994) were established whose main purpose were to promote organic farming in Nepal. Nepal Permaculture Group (NPG) was established in 1992 to act as an umbrella organization of sustainable farming, organic farming, and permaculture. It helped to network like-minded individuals and organizations. It currently has 19 organizational members and more than 1200 trained individual members who are advocating for policy formulation, research and trainings, and partnerships with government and international organizations. It played a lead role in developing national guidelines on organic farming.

On the policy side, though emphasis on chemical fertilizers to improve agricultural production is more evident, the government support for farming sustainability is also growing by highlighting topics such as integrated pest management, integrated soil nutrient management, promotion of agroforestry, sustainable rice intensification, and organic farming, among which organic farming is well accepted in terms of production and institutional development. The Agricultural Perspective Plan (APP) 1995 was the first consolidated and comprehensive long-term strategic plan for farming sector. Its main objective was to commercialize subsistence farming through the supply of fertilizer, irrigation, rural agricultural road construction, rural electrification, and the use of agroecologically appropriate technology to accelerate the economic growth. The only aspect it recognized that was close to organic farming was its emphasis on Integrated Pest Management (IPM) from the realization of harmful effects of chemical pesticides. The National Fertilizer Policy 2002 emphasized on balanced use of organic and inorganic fertilizers. It highlighted Integrated Plant Nutrient Management System (IPNMS) for efficient and balanced use of fertilizers. The National Coffee Policy 2003 was formulated realizing commercial value of organic coffee in Nepal and proposed development of organic coffee with a national logo. It was motivated by economic gain rather than environmental care and fairness. National Agricultural Policy 2004 encouraged organic farming through support for certification and accreditation; encouraged production, use, and promotion of organic fertilizers; and developed food standards to control quality and certify food products. Agribusiness Policy of 2006 proposed to establish organic/pesticides-free production area based on three categories: commercial crop/commodity production area, organic/pesticide-free production area, and agricultural products export area. Agricultural Biodiversity Policy 2006 authorized ban on importing genetically modified organisms (GMOs) that have potential risk on the environment and emphasized use of bio-fertilizers and bio-pesticides. Safe pesticide use, IPM and organic farming were prescribed for conserving pollinators and their habitats. Trade Policy 2009 provides assistance regarding packaging, storing, and certification of agricultural products to make them compelling in an international market. Among other major agricultural commodities, in terms of organic products, it supported tea, coffee, honey, and vegetables, including their certification for export. Nepal Trade Integration Strategy 2010 recommended implementation of a policy and institutional system for issuing an internationally recognized organic certificate developed by the government, especially for exportable products such as honey, tea, coffee, and medicinal herbs. Climate Change Policy 2011 has not mention organic farming despite its contribution in climate change mitigation and adaptation. Agricultural Development Strategy 2014 focused on increasing soil organic matter (SOM) from 1% to 2% within 5 years and 4% within 10 years. It continued to support organic/bio-fertilizer as supplementary and complementary to chemical fertilizers for sustaining soil fertility and to attain higher productivity and emphasized on branding of organic agricultural products for competitiveness and export.

The most significant achievement that laid the foundation for organic farming development is National Standards of Organic Agriculture Production and Processing 2007. It is a government-competent organic labeling regulation that provides rules, regulations, guidelines, and procedure for the production and processing of organic products and regulates, monitors, and evaluates organic farming development in Nepal. The standard can be found for land; conversion period; crop production; soil, water, and manure management; diseases, insects, and weed management; storage procedures; livestock farming; fish farming; bee farming; processing, packaging, and storage; and social responsibility and fair business. It prohibits chemical contamination in production, transfer, and processing and use of GMOs and radioactive devices. It emphasizes the use of local variety, organic seed source, and seed without chemical treatment. It however limits the use of synthetic fertilizers and contaminated manure.

This standard is also criticized for its lack of clarity though. For example, Article 3.4.4 allows use of 5–10 kg potassium fertilizer per metric ton of compost, but Article 3.4.8 limits its use to 5 kg per metric ton of compost. Likewise, Article 3.5.6 allows use of chemical pesticides for pest control. The certification standard does not consider local and regional differences and the cost of certification is only affordable by large farms. There is no differentiation for smallholders who prefer to go only for domestic market. So far seven international organic standards and regulations exist in Nepal: United States Department of Agriculture (USDA)'s National Organic Program (NOP); European Commission (EC); Food and Agriculture Organization of the United Nations/World Health Organization (FAO/WHO) Codex Alimentarius Commission (CAC); Japanese Agricultural Standard (JAS); International Federation of Organic Agriculture Movements (IFOAM); National Organic Standard, India; and Department of Agriculture, Australia. However, such third-party certification (TPC) is unaffordable to smallholder farmers. As an alternative Internal Control System (ICS) was developed that was endorsed on September 14, 2012, by the government of Nepal. A third-party certification bodies inspect the proper functioning of the system and perform few spot-check reinspections of individual smallholders within a group. This certification system needs help of external third-party certification body that audits the group's ICS, which proved to be still costly for the poor farmers. Thus, came the concept of Participatory Guarantee Systems (PGS) that was endorsed on February 19, 2013, by the government of Nepal. PGS is a locally focused quality assurance system that is based on stakeholders' active participation and is built on the foundation of trust, social networks, and knowledge exchange. It is especially designed for the local market but problems in record keeping and lack of trust and institutional networks among producers and consumers caused PGS to be almost nonexistent in practice in Nepal.

Regarding these, some working guidelines/procedures have also been established by the government for the promotion of organic farming, such as Working Procedure for Monitoring Organic and Bio-fertilizer 2068, Procedure for Registration Process on National Accreditation Body for Organic Agriculture 2069, Guidelines for Group Certification Internal Control System of Organic Agriculture Production 2069, Guidelines for Participatory Guarantee System 2069, Working Guidelines for Subsidies on Organic Agriculture Certification Process for Export 2069, Working Procedures for Subsidies for Implementing Internal Control System of Organic Certification 2071, and Working Procedures for additional budget allocated for Village Development Committees (VDCs, lowest administrative unit) for organic agriculture 2071 (Atreya 2015).

Overall, policy implementation is far below what has been stated theoretically. Lack of adequate and integrated research, extension, manpower and other support on organic farming production, marketing, and input supply have hindered the development of organic farming promotion. Organic product legislation, standardization, certification, and infrastructure development are also the major concerns. Basically the policy provisions are too broad without clear pathways to translate them into actions (Pandey 2012). On the optimistic side, the growing support of government to uplift this sector looks very promising. After few years of informal piecemeal support, the government has finally incorporated promotion of organic sector in its Agricultural Perspective Plan. Its Agricultural Development Strategy 2015 now includes support for organic sector development such as subsidy to establish organic fertilizer factories, subsidy of Nepalese rupees (NRs.) 10/kg for purchase of organic fertilizer, support for cattle housing improvement in 60 out of 75 districts, allocation of 25% budget to those VDCs adopting "organic pockets" within their villages, and continuation of subsidy for certification cost for export market and for establishing ICS (Willer and Lernoud 2016).

2.6 Summary

Agriculture is one of the major sectors in Nepal contributing significantly to its economy and absorbing the largest labor force. Government support to uplift this sector is more inclined toward improving conventional farming. However, impact of such farming system is now evident in areas where use of chemicals is higher through environmental degradation and lower soil quality. Declining soil fertility, negative repercussions on environment and health of farmers due to use of agrochemicals, and market demand reinforced organic movement in Nepal. In the midst of all these, Nepalese farming sector has to tackle the issue of climate change and food insecurity and satisfy the food requirement of current and future population that is expected to grow. The importance of organic farming on these issues should not be overlooked as well. Increasing awareness on these issues has finally led government to include organic farming-friendly policies. The most important standard that laid the foundation for organic farming development is National Standards of Organic Agriculture Production and Processing 2007, although it contains certain drawbacks in terms of clarity. In its Agricultural Development Strategy 2015, it has pledged to provide subsidy for organic fertilizer, certification cost for export market and establishing Internal Control System, support cattle housing improvement, and develop "organic pockets" within villages. Overall, policy implementation is far below what has been stated theoretically. Lack of adequate and integrated research, extension, manpower and other support on organic farming production, marketing, and input supply have hindered the development of organic farming. Organic product legislation, standardization, certification, and infrastructure development are also the major concerns. Basically the policy provisions are too broad without clear pathways to translate them into actions.

References

Adhikari RK (2011) Economics of organic rice production. J Agricult Environ 12:97–103
Atreya K (2015) In search of sustainable agriculture: a review of national policies relating to organic agriculture in Nepal. Asia Network for Sustainable Agriculture and Bioresources (ANSAB), Kathmandu
Bhandari DR (2006) Community level organic vegetable production program: an experience of Kathmandu district. In: Proceedings of a first national workshop on organic farming. Directorate of Agriculture Extension (DoAE), Lalitpur
Bhatta GD (2010) Stakeholder and spatial perspectives of organic farming in Nepal. LAP LAMBERT Academic Publishing, Deutschland
Bhatta GD, Doppler W (2010) Socio-economic and environmental aspects of farming practices in the peri-urban hinterlands of Nepal. J Agricult Environ 11:26–39
Bhattarai M (2006) Nepal. Keio University, Tokyo
Bishwajit G, Sarker S, Kpoghomou MA, Gao H, Jun L, Yin D, Ghosh S (2013) Self-sufficiency in rice and food security: a South Asian perspective. Agricult Food Security 2:10
Brandt K (2007) Organic agriculture and food utilization. Newcastle University, United Kingdom
CBS (2011) Preliminary result of national population census 2011. National Planning Commission Secretariat, Central Bureau of Statistics, Kathmandu
CBS (2013) Nepal thematic report on food security and nutrition. National Planning Commission, Central Bureau of Statistics (CBS), Kathmandu
DoAE (2006) Proceedings of a first national workshop on organic farming. Directorate of Agriculture Extension (DoAE), Lalitpur
FAO (2006) Food security. Agriculture and Development Economics Division (ESA), Food and Agriculture Organization of the United Nations (FAO), Rome
FAO (2008) Climate change and food security: a framework document. Food and Agriculture Organization of the United Nations, Rome
Ghimire S, Dhungana SM, Vijesh VK, Teufel N, Sherchan DP (2012) Biophysical and socioeconomic characterization of the cereal production systems of Central Nepal. Socioeconomics Program Working Paper 9, International Maize and Wheat Improvement Center (CIMMYT), Mexico

References

IFOAM (2009) High sequestration low emission food secure farming. International Federation of Organic Agriculture Movements, Bonn

Karki YK (2015) Nepal Portfolio Performance Review (NPPR). Ministry of Agricultural Development (MoAD), Kathmandu

KC GK (2006) An idea on organic agriculture system in Nepal. Proceedings of a first national workshop on organic farming. Directorate of Agriculture Extension, Kathmandu, pp 10–26

Kinyangi J (2007) Soil health and soil quality: a review. http://www.worldaginfo.org/files/Soil%20Health%20Review.pdf. Retrieved 7 Nov 2011

Koirala P, Dhakal S, Tamrakar AS (2009) Pesticide application and food safety issue in Nepal. J Agricul Environ 10:111–114

Mallick V (2012) Policy and regulatory requirements for agricultural commercialization in Nepal. Ministry of Agriculture and Cooperatives (MoAC), Kathmandu

MoAD (2015) Ministry of Agriculture Development (MoAD). .Welcome to Department of Agriculture. http://www.doanepal.gov.np/. Retrieved 17 Jan 2015

MoE (2011) Status of climate change in Nepal. Ministry of Environment, Government of Nepal, Kathmandu

Niggli U, Schmid H, Fliessbach A (2007) Organic farming and climate change. Research Institute of Organic Agriculture (FiBL), Frick

Palikhe B (2002) Challenges and options of pesticide use: in the context of Nepal. Landschaftsökologie und Umweltforschung 38:130–141

Pandey B (2012) State policy and institutional arrangement on organic agriculture promotion. Proceedings of the national workshop on organic agriculture-practice to policy, Nepal Permaculture Group (NPG), Lalitpur, pp 24–27

Pant KP (2006) Policies and strategies of Nepal government to promote organic farming in the context of Nepal's membership to WTO. Proceedings of a first national workshop on organic farming. Directorate of Agriculture Extension (DoAE), Lalitpur

Paudel MN (2016) Prospects and limitations of agriculture industrialization in Nepal. Agronom J Nepal 4:38–63

Pokhrel DM, Pant KP (2009) Perspective of organic agriculture and policy concerns in Nepal. J Agricult Environ 10:89–99

Rahija M, Shrestha HK, Stads GJ (2011) Nepal: recent developments in public agricultural research. Agricultural Science & Technology Indicators (ASTI), Kathmandu

Regmi HR (2007) Effect of unusual weather on cereal crops production and household food security. J Agricult Environ 8:20–29

Samie A, Abedullah AM, Kouser S (2010) Economics of conventional and partial organic farming systems and implications for resource utilization in Punjab (Pakistan). Pak Econom Soc Rev 48 (2):245–260

Samriddhi (2011) Commercialization of agriculture in Nepal. Samriddhi – The prosperity foundation, Kathmandu

Scialabba NEH (2007) Organic agriculture and food security. International conference on organic agriculture and food security. Food and Agriculture Organization of the United Nations (FAO), Rome

SECARD (2011) Market oriented organic agriculture promotion project in Chitwan district of Nepal. SECARD, Kathmandu

Shah BP, Devkota B (2009) Obsolete pesticides: their environmental and human health hazards. J Agricult Environ 10:51–56

Shrestha PL, Neupane FP (2002) Socio-economic contexts on pesticide use in Nepal. Landschaftsökologie und Umweltforschung 38:205–223

Shrestha RK (2010) Fertilizer policy development in Nepal. J Agricult Environ 11:126–137

Sligh M, Christman C (2007) Organic agriculture and access to food. International conference on organic agriculture and food security. Food and Agriculture Organization of the United Nations (FAO), Rome

Smith P, Martino D, Cai Z, Gwary D, Janzen H, Kumar P, McCarl B, Ogle S, O'Mara F, Rice C, Scholes B, Sirotenko O (2007) Agriculture. In: Metz B, Davidson OR, Bosch PR, Dave R, Meyer LA (eds) Climate change 2007: contribution of working group III to the fourth assessment report of the intergovernmental panel on climate change. Cambridge University Press, Cambridge/New York

Tamang S, Dhital M, Acharya U (2011) Status and scope of organic agriculture in Nepal. Food and Sustainable Agriculture Initiative. ForestAction, Lalitpur

Vaidya GS, Shrestha K, Wallander H (2008) Effect of plant residue on AM fungi. Scientific World 6(6):85–88

Weiss J (2004) Global organics. http://newhope360.com/agriculture/global-organics. Retrieved 16 Mar 2012

WFP (2009) .The future of food: Creating sustainable communities through climate adaptation. http://reliefweb.int/sites/reliefweb.int/files/resources/F1AA71E084C1D47B852576BD00634D1A-Full_Report.pdf. Retrieved 10 Dec 2011

Willer H, Kilcher L (2009) The world of organic agriculture: statistics and emerging trends. International Federation of Organic Agriculture Movements (IFOAM)/Research Institute of Organic Agriculture (FiBL), Bonn/Frick

Willer H, Kilcher L (2010) The world of organic agriculture: statistics and emerging trends. International Federation of Organic Agriculture Movements (IFOAM)/Research Institute of Organic Agriculture (FiBL), Bonn/Frick

Willer H, Kilcher L (2012) The world of organic agriculture: statistics and emerging trends. International Federation of Organic Agriculture Movements (IFOAM)/Research Institute of Organic Agriculture (FiBL), Bonn/Frick

Willer H, Lernoud J (2016) The world of organic agriculture: statistics and emerging trends. International Federation of Organic Agriculture Movements (IFOAM)/Research Institute of Organic Agriculture (FiBL), Bonn/Frick

Willer H, Yussefi M (2006) The world of organic agriculture: statistics and emerging trends. International Federation of Organic Agriculture Movements (IFOAM)/Research Institute of Organic Agriculture (FiBL), Bonn/Frick

World Bank (2015) Nepal: country overview. http://www.worldbank.org/en/country/nepal/overview. Retrieved 20 July 2015

Yussefi M, Willer H (2003) The world of organic agriculture: statistics and emerging trends. International Federation of Organic Agriculture Movements (IFOAM)/Research Institute of Organic Agriculture (FiBL), Bonn/Frick

Zundel C, Kilcher L (2007) Organic agriculture and food availability. Research Institute for Organic Agriculture (FiBL), Frick

Chapter 3
Organic Farming in Chitwan District of Nepal

Abstract The use of agrochemicals and pesticides is very much common in Chitwan District of Nepal, but group conversion to organic farming also exists. Group formation has led to higher adoption rate of organic farming, although a closer scrutiny shows that being a member of such group does not guarantee that all farmers will undeniably practice organic farming over the years. This chapter analyzes differences in households practicing organic and conventional farming in terms of their socioeconomic background and the functioning of the group itself that has led some farmers to divert their practice back to conventional farming. While group formation plays a crucial role in commencing organic farming, it comes with numerous challenges of unequal distribution of assistances, unequal participation in different group activities, less saving interest rate, and unequal advantage of premium market. Among others, establishing link within untapped local organic market can be one way to revitalize organic farming in the study area.

3.1 Introduction

This chapter discusses the status of organic farming in Chitwan District of Nepal. Geographically, Nepal is divided into three ecological zones and five development regions. Chitwan District lies in southern part of Central Development Region of Nepal. The southern part is basically a plain area, also known as Tarai region (Fig. 3.1). It accounts for 20.1% of total land area, but 34% of total cultivable land area of the country lies in this part as it has the most fertile soil compared to other parts of the country (FAO 2013). In fact, it is also known as "the food bowl" of the country because of its subtropical climate that allows for the practice of commercial farming production (Paudel 2016). This lowland Tarai region produces an agricultural surplus, portion of which is supplied to other parts of the country (SECARD 2011). But it is not immune from food deficiency as has been reported for the year 2013/2014 because of unpredictable weather and high weather-dependence nature of farming (Paudel 2016).

The reason Chitwan District was selected for the case study is because agriculture is the primary source of income for majority of the population and it is also one of the main commercial cultivation areas with a high volume of agrochemical usage (Neupane et al. 2014). It lies between 27° 21′ 45″ to 27° 52′ 30″ north latitude and

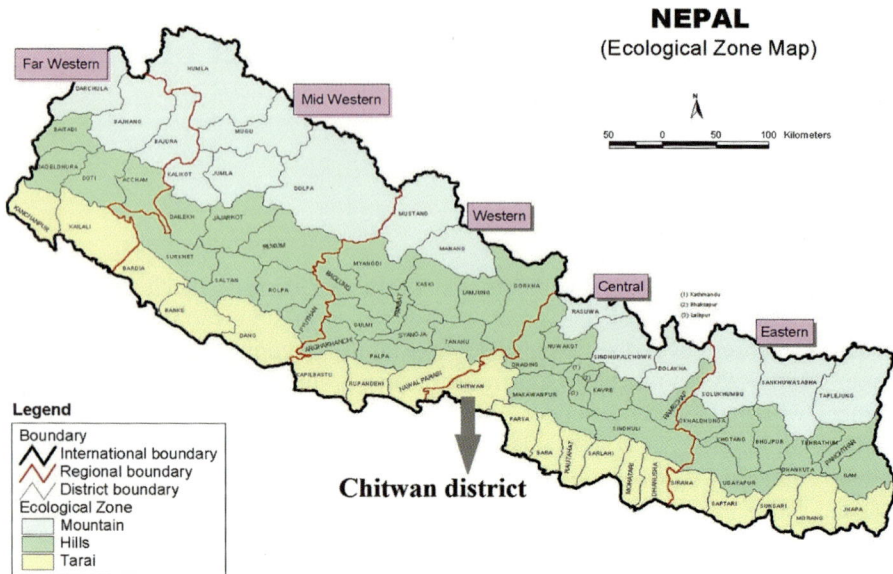

Fig. 3.1 Map of Nepal showing ecological zones, regional boundaries, and study area (*Source*: http://huebler.blogspot.jp/2015_05_01_archive.html)

83° 54′ 45″ to 84° 48′ 15″ east longitude. It occupies an area of 2205.9 km² (Devkota et al. 2011). It is bordered by Makwanpur and Parsa districts on the east, Nawalparasi and Tanahun districts on the west, Gorkha and Dhading districts on the north, and Indian state of Bihar and Uttar Pradesh on the south. Its elevation rises from 144 masl to 1947 masl (DDC 2017). The population of this district is 566,661, comprising almost 12% of the population in central Tarai region and 2.13% of the whole country. The average household size is 4.27, which is low compared to 5.46 of central Tarai region and 4.7 of Nepal. The population growth rate of 20.04% from 2001 to 2011 was fairly higher compared to the country average of 14.99% (MoAD 2014).

Chitwan District has a climate of subtropical monsoon with an average annual rainfall of 2318 millimeters (mm). It has high agricultural potential with the most fertile alluvial floodplain land, forest, rivers, and lakes in Nepal (Devkota et al. 2011). According to MoHP (2011), only 27% of the population in Chitwan District resides in urban area, which means that the rest 73% live in rural area where agriculture is the mainstay. Various crops are grown and livestock reared in this district (Table 3.1). Compared to the productivity of central Tarai region and Nepal as a whole, it fares low in terms of spices, pulses, oil seeds, and cash crops but has higher productivity in terms of citrus fruit, meat, egg, wool, and fish. Finally its productivity is higher than national average but lower than central Tarai region's productivity in terms of cereals, vegetables, tropical fruits, and milk (MoAD 2014). The use of chemical fertilizer has been increasing quite drastically since 2010/2011

3.1 Introduction

Table 3.1 Crops, livestock, and fishery productivity comparison in Chitwan District, central Tarai region, and Nepal

Crop/livestock categories (production unit)	Chitwan District	Central Tarai region	Nepal	Remarks
Cereals (kg/ha)	11,992	30,865	10,705	Paddy, maize, millet, wheat, barley
Vegetables (kg/ha)	13,748	13,755	13,419	Cauliflower, cabbage, broccoli, tomato, radish, broad mustard leaf, carrot, turnip, capsicum, peas, French beans, French beans (pole type), French beans (bush type), French beans (sword type), broad beans, asparagus beans, cowpea, other legumes, asparagus, tree tomato, Akabare chilli, chilli, okra, brinjal, onion, cucumber, pumpkin, squash, bitter gourd, pointed gourd, sponge gourd, ridge gourd, snake gourd, bottle gourd, ash gourd, balsam gourd, Kakari, Kundru, chayote, watermelon, other cucurbits, drumsticks, lettuce, fennel leaf, coriander leaf, spinach, cress, *Amaranthus*, fenugreek leaf, Swiss chard, other leafy vegetables, *Colocasia*, yam, elephant foot yam (Ol), other tubers, other vegetables
Spices (mt/ha)	21.9	38.78	31.95	Ginger, garlic, turmeric, chilli
Pulses (kg/ha)	7986	8092	8296	Lentil, chickpea, pigeon pea, black gram, grass pea, horse gram, soybean, others
Oil seeds (kg/ha)	3990	4603	5243	Mustard, sarsoon, rayo, sesame, linseed, niger
Citrus fruit (mt/ha)	24.3	–	24.12	Mandarin, lime, lemon
Tropical fruit (mt/ha)	11.52	14.48	9.18	Mango, banana, guava, papaya, jackfruit, pineapple, litchi, areca nut, coconut
Cash crop (kg/ha)	48,686	58,634	57,703	Oil seed, sugarcane, potato
Milk (mt/livestock)	0.8	0.84	0.72	From milking cow and buffalo
Meat (mt/livestock)	0.094	0.071	0.060	From buffalo, sheep, goat, pig, fowl (including chicken), duck
Egg (no. in '000/livestock)	0.13	0.12	0.1	Egg-laying hen and duck
Wool (kg/sheep)	2.59	2.53	0.74	
Fish (kg/ha)	4764	4656	4352	

Source: MoAD (2014)

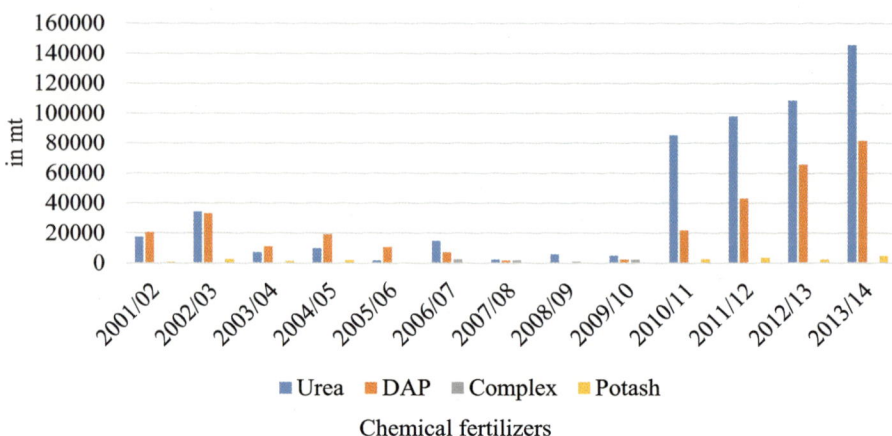

Fig. 3.2 Annual sales of chemical fertilizer in Nepal 1992/1993–2013/2014 (*Source*: MoAD 2014)

in Nepal, among which urea, diammonium phosphate (DAP), and potash are most significantly used (Fig. 3.2). In 2013/2014, 145,622 mt of urea, 81,520 mt of DAP, and 5046 mt of potash were sold all over the country. Out of this, Chitwan District used 3.9% (5679 mt), 3.24% (2645 mt), and 10.64% (537 mt) of urea, DAP, and potash, respectively. In aggregate, it used 3.82% (8861 mt) of chemical fertilizer that was sold throughout Nepal (MoAD 2014).

Indiscriminate use of agrochemicals in Chitwan District is very much existent, resulting in pest resistance toward pesticides, resurgence of new or already eradicated diseases and pests, and other health hazards to people that are not studied well yet (SECARD 2011). There are few studies that have highlighted the drawbacks of using such chemicals particularly in Chitwan District. Pudasaini et al. (2016) report that the heavy use of chemical pesticides wiped out all domesticated bees and destroyed many colonies. The bee pollinators of mustard were affected by the high-level usage of pesticides to the crops. Although farmers did agree that the natural pollinator population was decreasing, they were not aware of the negative effect of chemical pesticide on insect pollinators. They also attributed to the imbalanced use of mono-fertilizer (urea) and less use of farm yard manure (FYM) and other nutrient-containing fertilizers for lower production of rapeseed. A study by Ghimire and Khatiwada (2001) reported on a particular city of eastern Chitwan where more than 50% of pesticides is either misused or overused, thus creating a huge possibility of damaging nontarget groups such as humans, pet animals, natural enemies, and the environment. The pesticides are used haphazardly when farmers do not even wait for a prescribed period of 10 days to 1 month (depending on the pesticide type) after the application of pesticides but rather sell immediately after, which can have negative repercussions on the health of consumers. Toxic pesticides are used just to brighten the color of vegetables to make it appealing to the consumers. Farmers do not consider the prescribed dose of pesticides to be used and were found

3.1 Introduction

applying repetitively, without using any protective gears such as apron, mask, etc., and spraying them in the bright sunlight. They were also found to be using highly toxic pesticides already banned in developed countries. It is understood that more than 90% of the applied pesticides ultimately goes to the soil, which in turn can destroy the soil microfauna. Excess pesticide use has amplified pest problem that incurs heavy losses in major vegetables. Pests like diamondback moth (*Plutella xylostella*) and cabbage butterfly (*Pieris brassicae*) can be found in cole crops, brinjal fruit and shoot borer (*Leucinodes orbonalis*) in brinjal, red pumpkin beetle (*Aulacophora foveicollis*) in pumpkin, cucurbit fruit fly (*Dacus cucurbitae*) in summer squash, aphid (*Myzus persicae*) in radish, tomato fruit borer (*Helicoverpa armigera*) in tomato, soybean hairy caterpillar (*Spilarctia casigneta*) in legumes, and aphid (*Acyrthosiphon pisum*) in beans. Despite its huge agricultural potential and using production-enhancing chemical inputs, in 2013/2014 Chitwan District had food requirement deficit of 55,332 mt, while central Tarai region and Nepal as a whole had surplus of 17,583 mt and 789,890 mt, respectively (Table 3.2) (MoAD 2014).

However, in some areas the concept of organic farming has also been emerging, especially in three adjoining VDCs, namely, Phoolbari, Shivanagar, and Mangalpur VDC (Fig. 3.3). Previously in Phoolbari VDC, some enthusiastic farmers for reasons of health, soil fertility, to avoid increasing cost of chemical inputs, or simply as a continuation of traditional farming started practicing organic farming. Overtime, more neighboring farmers joined in and was followed by establishment of informal farmers' group under District Agriculture Development Office (DADO) to get various agricultural assistances from the government. DADOs have been established in all 75 districts of the country for agriculture-related extension services to agricultural producers. It performs functions of information dissemination on improved farming techniques through the use of various extension methods including demonstration, training, farm visit, farm tour, competition, leaflet distribution, and meeting (FAO 2010). Later the group was renamed to reflect its purpose of uplifting organic farming and soon converted into a cooperative named Organic Agriculture Producers' Cooperative Limited. Amid this transition, some NGOs (SECARD Nepal, Eco Centre, and Action Aid) initiated various organic farming-related projects, and it was expanded in two other adjoining VDCs as well. Shivanagar VDC consists of one group, whereas in Mangalpur VDC, three groups have been formed. All groups in these two VDCs are informal in nature. Farmers have been receiving training related to organic farming from general to more specific ones such as preparation of bio-fertilizers and pesticides for insect/pest management, market promotion, and network development; distributing pamphlets on Plant Health and Clinic Initiative; setting up hoarding boards for raising awareness; developing resource center; operating Farmer's Field School (FFS); technology development and transfer; and other extension services (SECARD 2011). Due to its vitality, it has also garnered interest from various individuals researching on organic farming, and accordingly a number of studies have been conducted on one or more of these VDCs (Adhikari 2009; Adhikari 2011; Bhat and Ghimire 2008; Organiconepal 2006; Kafle 2011a, b). It is also quite popular for homestays where

Table 3.2 Food availability and requirement in 2013/2014 in Chitwan District, central Tarai region, and Nepal

Food availability and requirement (in mt)	Chitwan District	Central Tarai region	Nepal
Edible production	52,804	926,117	6,085,776
Requirement	108,136	908,534	5,295,886
Balance	−55,332	17,583	789,890

Source: MoAD (2014)
Note: The above data is based on six cereal (staple) crops such as rice, maize, millet, buckwheat, wheat, and barley

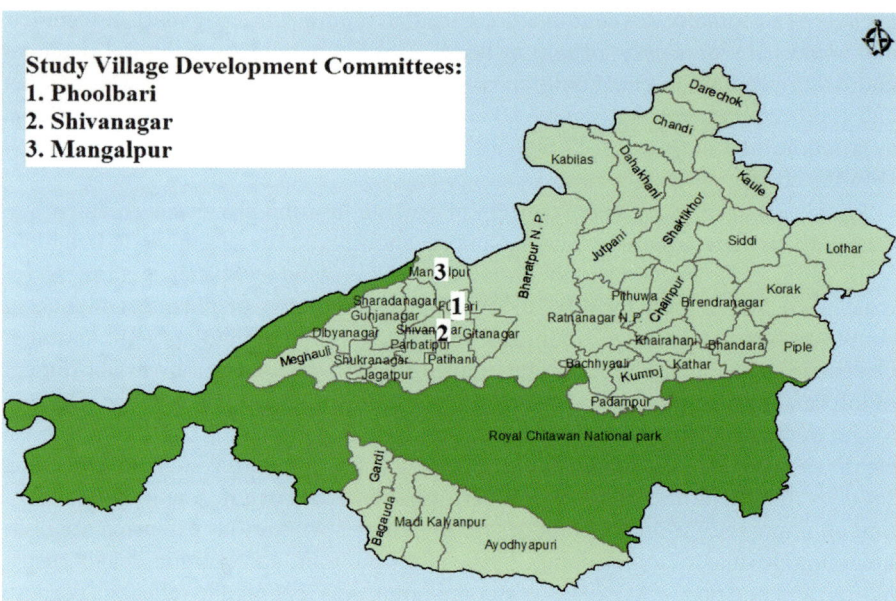

Fig. 3.3 Map of Chitwan District showing study areas (VDCs) (*Source*: http://neksap.org.np/allmaps/fsp-mid-nov13-to-mid-march14-chitwan)

students, foreigners, and other enthusiasts come to learn about organic farming firsthand. Thus, these three VDCs were chosen as study areas.

3.2 Source of Data

The study uses both primary and secondary data collected through individual household survey using semi-structured questionnaire, researcher's observation, and participatory methods such as focal group discussions and key-informant interviews. Field visits for data collection were done in two phases. First phase of the survey to collect household data using small-scale sample survey and to observe firsthand the status quo of organic farming was done from February 2013 to March 2013. Follow-up survey to gather additional information through participatory methods such as focal group discussions and key-informant interviews was done from October 2014 to November 2014.

3.3 Sample Design

A sample of 300 individual households (initially to choose equal number of organic and conventional farmers) were selected using stratified random sampling method by taking members of a group formed for the purpose of organic farming and nonmembers as strata and were interviewed using semi-structured questionnaire. Currently a cooperative in Phoolbari VDC has 125 members, an informal group in Shivanagar VDC has 44 members, and Mangalpur VDC has a total of 90 members with 30 members in each of the three informal groups. Members of such formed groups thus became the potential respondents under the hypothesis that all farmers belonging to such group would be organic farmers. However, during field survey, it was realized that not all farmers belonging to such group are actually practicing organic completely. Likewise, some nonmember farmers were also found to be practicing organic farming, which means that organic farming for them is a way of giving continuation to what their forefathers practiced. Thus, in the sample, both organic and conventional farmers cannot simply be identified from their membership status in such groups.

There are various requirements laid down by the "National Standards of Organic Agriculture Production and Processing 2007" as guiding principles for practicing organic farming. Among others, in order to prevent contamination from conventional farm/products in an organic farm, there should be a buffer zone. The crops grown in such zone cannot be considered as organic. If in case there is a road in between these two farms, a minimum of 4 m distance should be maintained, and 5 m if these farms happen to be adjacent to each other. Similarly, chemical fertilizers and pesticides are prohibited; and all machineries and tools should be properly cleaned if they had been used for conventional farming purpose, before it can be used for organic farming (MoAC 2008). However, such rigorous practice of following all conditions was found to be absent and understandably so because it would take enormous resources from these smallholder farmers. For example, while it was possible to clean the machinery before milling the husked (organic) rice, the practice of using buffer zone and treating crops grown on such area as conventional was nonexistent. In fact, most of the farms did not have any specific technique of protecting one's organic farm from the conventional farm. Usually these two farms would be divided by a ridge of about 5–10 in. (Fig. 3.4). Thus, for the purpose of simplicity, this study implies organic farming as the one that completely excludes the use of agrochemicals.

On the other hand, when viewing the nature of conventional farmers, it should be understood that they too do not use chemical fertilizers and pesticides in all of the

Fig. 3.4 Organic and conventional farms without buffer zone in between (*Source*: Field survey 2013)

crops but on the basis of crop requirement. The reasons farmers relied on such inputs are to contain pests, diseases, and weeds and to increase the production. The most common pattern observed was the segregation of farmland for organic purpose. This is generally true for vegetable farming which farmers grow organically only for home consumption and is mainly done on a small portion of their land (Fig. 3.5) but use chemical fertilizers and pesticides on cereal crops which are rather produced on a larger area. For some it was difficult to grow certain crops without the use of pesticides. For example, most farmers faced the problem of late blight disease in potato for which using pesticide was inevitable (Fig. 3.6). Other such crops where most conventional farmers used pesticides are tomato, kidney bean, spinach, cowpea, and mustard greens. They used chemical fertilizer mainly for cereal crops like rice, maize, wheat, and oat and even for oil seed crop such as mustard (Fig. 3.7). For others, they choose to grow commercially viable crops like carrot through conventional means for easier management of weeds and pests, as well as to intensify production. Basically there is no difference in the kind of crops grown by organic and conventional farmers. Organic farmers could grow similar crops without chemical input because they either grow on a small scale or are willing to let go of the damaged crops. Some claimed that crop variety also matters, for example, in case of potato and rice, some had access to a variety that is free from pest problem. However, one significant characteristic that could be pointed out about organic farmers is their willingness to farm organically despite the challenges and possible risks.

An effort was made to select equal number of respondents taking into consideration group membership, but because of limited number of organic farmers itself, the final sample out of the total 300 respondents came out as 32% organic farmers and the rest 68% as conventional farmers. Similarly, 50% of the respondents are from Phoolbari VDC, and the rest 17% and 32% are from Shivanagar and

Fig. 3.5 Small-scale organic vegetable farming usually for home consumption only (*Source*: Field survey 2013)

Fig. 3.6 Pesticide used for pests in potato (*Source*: Field survey 2013)

Mangalpur VDCs, respectively (Table 3.3). This sample is based on the number of group members in a particular VDC (48%, 17%, and 35% of group members belonging to Phoolbari, Shivanagar, and Mangalpur VDCs, respectively). Altogether, 47% of the respondents belonged to a group and the rest 53% did not. Respondents who were interviewed from outside such group were selected randomly based on close geographical proximity with those respondents belonging to a group. Across the VDCs too, sample size for member farmers in Phoolbari VDC is higher compared to the other two VDCs (Table 3.4).

Fig. 3.7 Farmer preparing for chemical fertilizer application (*Source*: Field survey 2013)

Table 3.3 Distribution of respondents belonging to two farming systems across VDCs and based on group membership

Variables	Farming systems		Total ($n = 300$)
	Organic ($n = 95$)	Conventional ($n = 205$)	
VDCs			
Phoolbari	64 (67.37)	87 (42.44)	151 (50.33)
Shivanagar	15 (15.79)	37 (18.05)	52 (17.33)
Mangalpur	16 (16.84)	81 (39.51)	97 (32.33)
Membership			
Yes	71 (74.74)	69 (33.66)	140 (46.67)
No	24 (25.26)	136 (66.34)	160 (53.33)

Source: Field survey (2013)
Note: Figures in parenthesis indicate column percentage

Table 3.4 Distribution of respondents across VDCs and based on group membership

Membership	VDCs			Total (n = 300)
	Phoolbari (n = 151)	Shivanagar (n = 52)	Mangalpur (n = 97)	
Yes	83 (54.97)	18 (34.62)	39 (40.21)	140 (46.67)
No	68 (45.03)	34 (65.38)	58 (59.79)	160 (53.33)

Source: Field survey (2013)
Note: Figures in parenthesis indicate column percentage

3.4 Households' Background

In this chapter, data was analyzed through descriptive statistical tools such as percentage, mean, and standard deviation. Tables 3.5 and 3.6 give descriptive analysis of 285 respondents' (after data cleaning) selected variables, which form the basis of various components of their socioeconomic background. Head of households (HHHs) are those who are responsible for making key decisions in the family matter. It is found that only 8% of households are female headed which is comprehensible as Nepalese society is mainly patriarchal-based. Average age of HHH is 50 years old. Some 7% of HHHs do not have any educational background. In other words, they are illiterate and cannot read or write at all. About 30% of them identified themselves as having only basic education, which is defined as those who can do simple reading and writing. Some 41% had education till secondary level (formal education of 10th grade and below), 11% had higher secondary education (formal education of 11th and 12th grade), 9% had bachelor's degree, and only 3% had master's level education. The average year of educational attainment of respondents is 7 years.

Majority (58%) of HHHs recognize farming as their primary occupation. Almost 79% of farmers owned the land they are farming, but 21% of them either rented in for kind or cash or mortgaged in farmland in addition to farming their own land. Farmers have been practicing organic farming for about 3 years, on average. However, since standard deviation (3.25 ± 7.48) is higher, it means that organic farmers are either traditional practitioners who have been following the same farming system their forefathers practiced or are new practitioners, most probably as a result of organic farming-related activities performed through the group. Labor force unit (LFU) is a standard unit of labor force which takes people aged 14–59, irrespective of their sex, as 1 and those below 14 and above 59 as 0.5. In this study, LFU excludes household member/s who have migrated, whether temporarily or permanently, and reflects only those who are available in the household. As a result, households have 4.28 LFU, on average. Likewise, livestock unit (LSU) is a standard way of measuring livestock holding which is calculated as an aggregate of different types of livestock kept at household in a standard unit by taking one adult buffalo as one LSU, one immature buffalo as 0.5 LSU, one cow as 0.8 LSU, one calf as 0.4 LSU, one pig as 0.3 LSU, one sheep or goat as 0.2 LSU, and one poultry as 0.1 LSU (CBS 2003). In the study areas, about 87% of households have livestock holding with 2 LSU, on average. There is a significant difference between two farming systems with most of the organic farmers holding higher livestock compared to conventional farmers.

Around 47% of respondents have membership in a group formed for the purpose of organic farming, and it is significantly different across the two farming systems. About 76% organic farmers are members compared to just 34% conventional farmers. This shows that being member of such group does not guarantee that a farmer will indefinitely practice organic farming, but it definitely represents having higher number of organic farmers. Nearly 44% of respondents have received

Table 3.5 Descriptive analysis of (categorical) variables across two farming systems

Variables	Farming system Organic (n = 91)	Conventional (n = 194)	Total (n = 285)	P-value
Gender of HHH				
Male	82 (90.11)	180 (92.78)	262 (91.93)	0.440
Female	9 (9.89)	14 (7.22)	23 (8.07)	
Education of HHH				
Illiterate	4 (4.40)	16 (8.25)	20 (7.02)	0.497
Basics	25 (27.47)	60 (30.93)	85 (29.82)	
Secondary and below	38 (41.76)	79 (40.72)	117 (41.05)	
Higher secondary	10 (10.99)	21 (10.82)	31 (10.88)	
Bachelors	10 (10.99)	15 (7.73)	25 (8.77)	
Masters	4 (4.40)	3 (1.55)	7 (2.46)	
Primary occupation of HHH				
Farming	50 (54.95)	114 (58.76)	164 (57.54)	0.543
Others	41 (45.05)	80 (41.24)	121 (42.46)	
Ownership				
Owned + lent in	20 (21.98)	41 (21.13)	61 (21.40)	0.871
Owned	71 (78.02)	153 (78.87)	224 (78.60)	
Livestock holding				
Yes	85 (93.41)	164 (84.54)	249 (87.37)	0.036**
No	6 (6.59)	30 (15.46)	36 (12.63)	
Membership				
Yes	69 (75.82)	66 (34.02)	135 (47.37)	0.000***
No	22 (24.18)	128 (65.98)	150 (52.63)	
Training				
Yes	70 (76.92)	55 (28.35)	125 (43.86)	0.000***
No	21 (23.08)	139 (71.65)	160 (56.14)	
Income source				
Farming only	40 (20.62)	20 (21.98)	60 (21.05)	0.793
Farming + non-farming	154 (79.38)	71 (78.02)	225 (78.95)	
VDC				
Phoolbari	61 (67.03)	80 (41.24)	141 (49.47)	0.000***
Others	30 (32.97)	114 (58.76)	144 (50.53)	
Credit				
Yes	9 (9.89)	18 (9.28)	27 (9.47)	0.869
No	82 (90.11)	176 (90.72)	258 (90.53)	
Final price				
Yes	22 (24.18)	44 (22.68)	66 (23.16)	0.780
No	69 (75.82)	150 (77.32)	219 (76.84)	
Selling crops in market				
Yes	70 (76.92)	155 (79.90)	225 (78.95)	0.566
No	21 (23.08)	39 (20.10)	60 (21.05)	

Source: Field survey (2013)
Note: Figures in parenthesis indicate column percentage
***1% and **at 5% level of significance

3.4 Households' Background

Table 3.6 Descriptive analysis of (continuous) variables across two farming systems

Variables (measurement)	Farming system								T-test (P-value)
	Organic (n = 91)			Conventional (n = 194)				Total mean ± SD	
	Min.	Max.	Mean ± SD	Min.	Max.	Mean ± SD			
Discrete									
Age (years)	30	72	48.23 ± 9.81	26	84	50.30 ± 12.26		49.64 ± 11.56	0.159
Education (years)	0	17	7.37 ± 5.58	0	17	6.43 ± 5.37		6.73 ± 5.45	0.177
Experience (years)	1	55	10.17 ± 10.25	–	–	–		3.25 ± 7.48	–
Training (times)	0	12	2.60 ± 2.66	0	8	0.56 ± 1.17		1.21 ± 2.02	0.000***
Continuous									
Labor (LFU)	1.5	10	4.28 ± 1.84	1	11	4.29 ± 1.85		4.28 ± 1.84	0.961
Livestock (LSU)	0	12.7	2.12 ± 1.67	0	13.7	1.85 ± 1.75		1.94 ± 1.72	0.219
Farm size (ha)	0.02	2.37	0.49 ± 0.39	0.01	2.7	0.51 ± 0.41		0.50 ± 0.40	0.633
Farm income (NRs./year)	1820	1,014,245	186,717 ± 170,360	2850	994,692	197,400 ± 186,134		193,989 ± 181,016	0.643
Non-farm income (NRs./year)	0	960,000	221,715 ± 220,117	0	1,080,000	192,911 ± 190,825		202,108 ± 200,702	0.259
Agrovet (km)	0.01	9	1.58 ± 1.34	0.01	15	1.79 ± 1.89		1.73 ± 1.73	0.323
Market (km)	0.04	15	3.07 ± 3.63	0.01	15	2.73 ± 3.30		2.84 ± 3.40	0.426
Commercialization (crop sold/produced)	0	4.76	0.66 ± 0.75	0	3.99	0.74 ± 0.66		0.71 ± 0.69	0.371
Cash income (NRs.)	0	226,648	48,999 ± 54,685	0	234,669	64,359 ± 59,887		59,455 ± 58,621	0.039**
Crop diversity (SHDI)	2.05	3.88	3.15 ± 0.38	1.72	3.81	3.11 ± 0.38		3.12 ± 0.38	0.349

Source: Field survey (2013)
Note: ***1%, **5%, and *at 10% level of significance

training related to organic farming. Like membership, it is also significantly different across the two farming systems, and receiving such training, while it does not ensure that a farmer will practice organic farming, again definitely has impact on larger share of farmers practicing organic farming. While most of such trainings are provided through such groups, some farmers received it from other sources as well. On an average, farmers have received such training related to organic farming at least once. On an average, farmers have 0.5 ha of operational farmland which means most farmers in study areas are smallholders. Some 21% of households relied solely on farming for their livelihood, while 79% have non-farm income as well from sources such as service, business, rent, remittance, pension, and wage laboring. Therefore, it can be implied that most households have other sources of income besides farming. Income from farming includes monetary and nonmonetary value of total production for past 1 year from farming of cereals, vegetables, spices, pulses, oil seeds, fruits, livestock products and by-products, and occasional income generated from selling trees as well. Income is calculated in Nepalese rupees, a monetary unit of Nepal (US$1 = NRs.98.56 (Source: Nepal Rastra Bank, March 31, 2013)). On an average, households have income worth NRs.193,989. Compared to farm income, households have higher non-farm income which stands at NRs.202,108, on average.

Around 50% of respondents are from Phoolbari VDC. There is a significant difference in number of respondents belonging to two groups of a farming system. Phoolbari VDC has higher number of organic farmers, while the other two VDCs (Shivanagar and Mangalpur) combined have higher number of conventional farmer respondents. Agrovets are an exclusive store for agriculture-related products where products for farming inputs (seeds, fertilizers, pesticides, etc.) and livestock such as veterinary medicine could be found in addition to various equipment. This study takes distance to agrovet and market to understand how these facilities could have influenced the farming practice. Average distance to nearest agrovet and market is 1.73 km and 2.84 km, respectively. Only 10% of respondents have taken credit for the purpose of farming. Respondents used credit for investing in commercial crops, livestock rearing, and/or for irrigation purpose. Average commercialization rate is 0.71, which is calculated as a ratio of total quantity of crops sold to total produced.

Approximately 23% of respondents know the final price at which consumer buys their product, as they are involved in selling directly to the consumers in the local market or keep themselves informed through various links. The rest 77% sell their products to middlemen at a farm gate price and thus are not up to date with the retail price. Roughly 79% of respondents sell crops in the market, and the remaining 21% use their farming production for own household consumption only. Gross farm cash income (income generated as a result of selling crops in the market without deducting the cost of production) is NRs.59,455, on average. It is significantly different across the two farming systems, with conventional farming having higher gross farm cash income compared to organic farming. Finally Shannon Diversity Index (SHDI), which is one of the indicators of crop diversity (Zhang et al. 2012), is calculated by taking into consideration all crops produced by the household under six categories: cereals, vegetables, spices, pulses, oil seeds, and fruits, including the area each crop is grown in (Appendix II). Households have SHDI value of 3.12, on average.

3.5 Nature of Group Formation

Since group membership alone does not indicate farmers' adoption of organic farming, this section examines the reasons for it. Table 3.7 provides information on status of farming practice of group members after formation of the group and reasons for practicing it. More so, it indicates how successful such groups have been in changing their members' farming behavior more toward organic farming since its inception.

Some 73% of member farmers practicing organic farming said that they switched to organic farming after being a member of a group formed for the purpose of organic farming. They implied that they have become more health conscious through various trainings and interactions that occurred as a result of being a member. About 27% of member farmers said they have been practicing organic farming even before the group came into existence. It means that these particular groups of farmers have been following traditional ways of farming, which their forefathers had practiced before them. These farmers are health conscious too, but this method of farming is what they are used to and never felt like needing chemical inputs. Therefore, the primary reason for practicing organic farming undeniably remains health rather than for monetary benefit from the premium market. Only one farmer claimed that he faced declining soil fertility over time instead of increase in the use of agrochemicals. Farmers have become health conscious, and as a result some also tend to grow the same crop organically for home consumption and conventionally to sell in the market. Although some organic farmers did confess that they were compelled to use chemicals for cereals few years ago. When there is a pest and disease problem, they can let go of other minor crops, as it is grown in a small amount. But for cereal crops, which are grown on larger scale and constitute larger portion of food consumption, it is riskier for farmers if in case they are not

Table 3.7 Farming practice after group formation

Farming status of member farmers after group formation	Member farmers		Total ($n = 140$)
	Organic ($n = 71$)	Conventional ($n = 69$)	
Changed to organic farming through group	52 (73.24)	–	52 (37.14)
Same as before	19 (26.76)	26 (37.68)	45 (32.14)
Later changed to conventional farming	–	19 (27.54)	19 (13.57)
Organic farming area ≥ 0.0676 ha[a]	–	24 (34.78)	24 (17.14)

Source: Field survey (2013, 2014)
Note: Figures in parenthesis indicate column percentage
[a]Minimum land area required to be group member

able to harvest the crops. Another challenge is organic seed that needs to be dried hence increasing the cost, whereas in conventional method, the use of pesticide is enough. The path to organic farming was also gradual for some farmers, meaning that they decreased the use of chemicals gradually so as not to face the sudden loss.

Similarly, around 38% of member farmers said they never changed to organic farming after becoming a member and continued using chemical inputs like before. Some farmers opined that they tried cultivating organically but whenever there is pests attack or disease in crops, they could not help but use chemical pesticides for fear of losing crops, thus not being able to follow organic farming even for 1 whole year. Contrastingly, there are nearly 28% member farmers who tried organic farming for a short period such as for a season, few months, or a year. However, they returned back to using chemical inputs entailing that organic farming requires more effort in terms of preparing and applying more amount of FYM, weeding, and so forth. Most importantly, they too suggested that organic means of pests and disease management take longer time and oftentimes face risk of crop failure within that duration. In addition to that, having to increase production for a particular season also remains their topmost priority rather than sustainability in a long run. In other words, it can be said that organic farming is mostly acceptable to the extent that it does not affect their food security.

From the field survey, it was also realized that having access to organically feasible crop varieties also indicates the possibility of practicing organic farming. Farmers claimed that crops like potato, tomato, and kidney bean are difficult to produce organically because of fungus. Crops that usually require chemical pesticides are spinach, cowpea, mustard greens, and mustard. Off-seasonal vegetables are difficult to grow organically and are usually possible only with the application of chemical inputs. In cereal crops such as rice, maize, wheat, and oat, use of inputs like urea, DAP, and muriate of potash (MOP) is very common. In rice too, some varieties (like *Sabitri* rice) do not need urea, but since all farmers cannot have access to such seed variety, they are compelled to grow those that demand some amount of chemical fertilizer. Approximately 35% of member farmers have been practicing organic farming in less than or equal to 0.07 ha of land which is the minimum land area required to be under organic management to be a member of such group. Among these, some farmers started out with more area but later reduced it to this minimum size. Thus, there are member farmers who are practicing conventional farming and simultaneously following organic farming on the minimum required land area needed for being a group member. In fact some farmers are also practicing on lesser area than the minimum required. This shows easing of rules or not taking any firm precaution by the group against members having less than required area under organic farming.

It was found that there are farmers who want to remain being a member and then those who no longer desire to be associated with such group anymore. Being a group member has its own pros and cons. In addition to being more health conscious and getting training on various facets of organic farming, they

3.5 Nature of Group Formation

Fig. 3.8 Package of vegetable seeds received by member organic farmers who participated in the training (*Source*: Field survey 2013)

occasionally are entitled to receiving various partial or full assistances such as seeds (Fig. 3.8), bio-pesticide container, water pump, compost-shed (Fig. 3.9), tricycle with a carrier (Fig. 3.10), tractor, partial certification cost, etc. But on the other hand, there are various strains among members both within the groups and across VDCs as a result of unequal distribution of assistances, unequal participation in different group activities, less saving interest rate, and unequal advantage of premium market. For instance, all three VDCs received 20 water pumps (about a size of 2 in. each) for irrigation and 20 bio-pesticide containers/sprayers. These were very limited given the huge number of members. It was thus prioritized based on those active members who regularly participate in group activities. In Mangalpur and Shivanagar VDC, they received one set of water pump for each group (4 sets for total of 134 members in these two VDCs), and the rest was distributed to farmers in Phoolbari VDC (16 sets for 125 members). Providers also justify that they allocate such things based on the number of members in a particular group in each VDC. Cooperative in Phoolbari VDC has also received NRs.20,000 worth of cart to collect farm produces and to sell. In 2014, they received assistance of 50% of the cost of tractor, and the rest was paid by members. They also got NRs.400,000 through Organic Certification Nepal (OCN) as half of the cost required for certification, and the rest again was paid by the group members themselves. This shows that member farmers in Phoolbari VDC have received more assistances. Cooperation among members gets more difficult with increasing number of members. Because of this, since 2015, Shivanagar VDC formed two informal groups from the previous one, namely, Nawa Kiran Prangarik Krisak Samuha which has 30 members (with four males) and Jan Kalyan Prangarik Krishak Samuha which also has 30 members (with 10 males). The low number of males in all the groups is because males are usually employed in non-farm sector, have to travel more often

Fig. 3.9 Improved compost-shed with a partial financial assistance from an NGO (*Source*: Field survey 2013)

Fig. 3.10 Tricycle with a carrier provided by an NGO for collecting and selling organic crops (*Source*: Field survey 2013)

than their female counterparts, and thus are less interested in meetings and other activities that a group member is expected to take part in.

Cooperative in Phoolbari VDC is a central collection point where produces from all member farmers spreading across three VDCs are brought so as to sell collectively to other cities where premium market exists. Farmers are able to get premium ranging from 9 to 140%, but again it is not equally accessible by all of its members. Many of those excluded farmers have a common complain that they are not informed at all about the demand for organic crops to be sold in the premium market in other cities and at times when they are, it is often too late and is in smaller quantities, making it economically infeasible for them as transportation cost to take produces to the collection center gets higher. A farmer in Shivanagar VDC complained that there is no vehicle to transport produces to Phoolbari VDC. Those excluded member farmers opined that they are also not made aware of various assistances received either from government or I/NGOs when the actual distribution takes place. On the other hand, those farmers who are responsible for such distribution claim that many farmers are not active when it comes to participating in various meetings. The only time they show up is once a month when it is time to make payment for monthly saving, while some farmers choose to be absent even on such occasion and rather send their children on their behalf. Gathering for training or various meetings is an important time for farmers to strengthen their link. Thus, only those farmers who are regular are prioritized more while distributing such assistances as that too come in limited quantity. Again those farmers who are involved in the management claim that they are not sure whether all farmers are actually practicing organic farming or to what extent and on which crops. A farmer in Phoolbari VDC argued that demanded variety and quantity of crop in the premium market itself are very limited, making it impossible or at least unrealistic to equally include all farmers. Thus, they only ask those farmers who have been more regular in group activities as that is one of the indications of their commitment toward farming organically. The survey conducted in 2014 found out that some local outlets have begun to sell organic or environment-friendly products that are within a few kilometers distance from the study VDCs, but farmers so far are not aware of it.

In a cooperative in Phoolbari VDC and groups of other two VDCs too, saving and credit facility exists for its members. While taking credit is based on individual farmer's requirement or choice, saving is mandatory for a member. The credit interest rate is 1% (for nonmember local farmers, the rate is 2%) for loan of NRs.1000, payable within 6 months, and 0.7% for saving (i.e., NRs.1000 loan equals NRs.10 as interest payable and Rs.1000 saving equals NRs.7 as interest receivable). Sometimes interest payable can be more than 1%, depending on money required to pay as saving interest to its members. However, similar other local financial institutions that are providing saving and credit facilities give credit interest of 12% and saving interest of 8%. Thus, some farmers, who do not need loan but want to save for generating some interest, wish to be nonmember as they could have received higher interest rate through other formal sources. As such they are not even getting any assistance by being a member, unlike their few selected associates. But doing so comes with deducting certain amount of their membership fee as a penalty for terminating their status, and thus farmers hesitate to be a nonmember. Thus, occasional assistances encourage some farmers to stay within the group by attempting to farm at least minimum land area required by such group to be practiced as organic, while some farmers desire to be nonmember which too comes with its own challenges.

3.6 Summary

Large-scale adoption of organic farming is made possible through group formation in Chitwan District of Nepal through which farmers are provided training and other extension services. However, rigorous practice of organic farming in line with the national standard is not done extensively as it would take enormous resources. Thus, this chapter simply segregates organic and conventional farming as restriction and use of chemical fertilizers and pesticides, respectively. The closer analysis suggests that over time, some farmers go back to practicing conventional farming and some farmers not belonging to such group are also practicing organic farming, which for them is following a traditional farming that their predecessors used to practice. Conventional farming in this particular case does not imply that farmers rely on chemical fertilizers and pesticides extensively but occasionally in the event of pests, disease, or weed outbreak, especially in case of vegetables and widely in case of cereal crops which are grown in a larger area. In some cases crop variety is also to blame for not being able to be grown organically.

As far as socioeconomic background of farmers in the study area goes, it is mainly patriarchal-based, as is predominant in the case of Nepalese society. Most households have up to secondary level education, and many households consider farming as their primary occupation. While renting and mortgaging farmland are prevalent, most of the farm is owned by farmers themselves, and majority of them own livestock. Livestock holding, membership in a group formed for the purpose of organic farming, and training related to organic farming are significantly higher among organic farmers. Taking credit for farming purpose is not so common among farmers, while selling crops in the market is done by majority of the households, but only few are aware of the actual price consumers pay for it. Conventional farmers earn higher farm cash income than organic farmers; however, there is no significant difference when it comes to household heads' age, education, labor availability, livestock holding, farm size, overall farm valuation, non-farm income, distance to significant establishments like agrovet and market, commercialization rate, and crop diversity.

Analysis of group formed for the purpose of organic farming showed that some farmers got more health conscious after being a member of such group; as a result of which, some also tend to grow the same crop organically for home consumption and conventionally to sell in the market. While selling of organic produce through such group in the cities at premium price is very limited, occasional financial and material assistance encourages some farmers to stay within the group by attempting to farm at least minimum land area required by such group to be practiced as organic. On the other hand, there are also farmers who are practicing organic farming on lesser land area than the required minimum. This shows that such formed groups do not have capacity enough to check the activities of its members. Overall, forming such group could be an efficient tool to introduce organic farming on a larger scale. However, it also comes with various other challenges of unequal member participation in group activities, entitlement by each member to the benefit received in the form of various assistances, and accessibility to the premium market. Smaller interest rate from savings compared to other financial institutions has also diminished the appeal of remaining a member in such groups. While many challenges are difficult to overcome, one of the ways to improve such situation could be to facilitate in selling local organic produces in the specialized shops which have begun to thrive very recently in the local area that can hopefully accommodate selling the produces from more farmers.

References

Adhikari RK (2009) Economics of organic vs. inorganic carrot production in Nepal. J Agricult Environ 10:23–38

Adhikari RK (2011) Economics of organic rice production. J Agricult Environ 12:97–103

Bhat BR, Ghimire R (2008) Promotion of organic vegetable production through farmers' field school in Chitwan, Nepal. 16th IFOAM Organic World Congress, International Federation of Organic Agriculture Movement (IFOAM), Modena

CBS (2003) National sample census of agriculture Nepal, 2001/02. National Planning Commission Secretariat, Central Bureau of Statistics, Kathmandu

DDC (2017) Brief introduction: Chitwan district. District Development Committee (DDC) Office, Ministry of Federal Affairs and Local Development, the Government of Nepal, Chitwan

Devkota R, Budha PB, Gupta R (2011) Trematode cercariae infections in freshwater snails of Chitwan district, central Nepal. Himal J Sci 7(9):9–14

FAO (2010) Agricultural extension services delivery system in Nepal. Food and Agriculture Organization of the United Nations, Kathmandu

FAO (2013) Country data: Nepal. http://www.fao.org/nr/water/espim/country/nepal/index.stm. Retrieved 17 Aug 2013

Ghimire A, Khatiwada BP (2001) Use of pesticides in commercial vegetable cultivation in Tandi, Eastern Chitwan, Nepal during 2001. Institute of Agriculture and Animal Science (IAAS), Chitwan

Kafle B (2011a) Diffusion of uncertified organic vegetable farming among small farmers in Chitwan District, Nepal: a case of Phoolbari Village. Int J Agricult Res Rev 1(4):157–163

Kafle B (2011b) Factors affecting adoption of organic vegetable farming in Chitwan District, Nepal. World J Agricult Sci 7(5):604–606

MoAC (2008) National standards of organic agriculture production and processing, 2064. Ministry of Agriculture and Cooperatives (MoAC), Kathmandu

MoAD (2014) Statistical information on Nepalese agriculture 2013/2014. Agri Business Promotion and Statistics Division, Ministry of Agricultural Development (MoAD), Kathmandu

MoHP (2011) Nepal population report. Kathmandu, Nepal: Ministry of Health and Population; Government of Nepal (MoHP)

Neupane D, Jørs E, Brandt L (2014) Pesticide use, erythrocyte acetylcholinesterase level and self-reported acute intoxication symptoms among vegetable farmers in Nepal: a cross-sectional study. Environ Health. doi:10.1186/1476-069X-13-98

Organiconepal (2006) Linking organic farmers in incentive sharing mechanisms through promoting local marketing systems: An initiative from Himalayan kingdom of Nepal. Organic Agriculture Research and Production Co. (Organiconepal) Pvt. Lld., Chitwan

Paudel MN (2016) Prospects and limitations of agriculture industrialization in Nepal. Agron J Nepal 4:38–63

Pudasaini R, Thapa R, Tiwari S (2016) Farmers perception on effect of pesticide on insect pollinators at Padampur and Jutpani VDCs, Chitwan, Nepal. Int J Appl Sci Biotechnol 4 (1):64–66

SECARD (2011) Market oriented organic agriculture promotion project (MOAP) in Chitwan district of Nepal. Society for Environment Conservation and Agricultural Research and Development (SECARD) Nepal, Kathmandu

Zhang H, John R, Peng Z, Yuan J, Chu C, Du G, Zhou S (2012) The relationship between species richness and evenness in plant communities along a successional gradient: a study from sub-alpine meadows of the eastern Qinghai-Tibetan Plateau, China. PLoS One 7(11):e49024

Chapter 4
Socioeconomic Dimension of Farming System

Abstract Though organic farming is sustainable, its share remains just 0.2% of the overall agricultural land in Nepal. In any adoption studies of agricultural innovations, socioeconomic variables are considered as important as agroecological variables and farmers' perception. This chapter assesses impact of farm households' socioeconomic variables that encourages or discourages adoption of organic and conventional farming systems in Chitwan District of Nepal. Data of 285 farm households was analyzed using binary logistic model. The result shows that while group formation plays a crucial role in commencing organic farming, it does not guarantee adoption over time because of varying levels of motivation among members. But households receiving higher number of organic farming-related training have highly significant probability to continue practicing organic farming. The impact of longevity of group formation and vitality of training is also reflected in adoption rate of organic farming among different village development committees considered for this study. Likewise, commercially available organic fertilizers and pesticides could also be playing significant role in the adoption rate of organic farming.

4.1 Introduction

Though organic farming is a growing phenomenon, its share in the global context still remains minimal. As of 2014, global total share of organic to overall agricultural land including in-conversion areas remains only 1%, and Nepal shares much smaller rate, which stands at 0.2% (Willer and Yussefi 2006; Willer and Lernoud 2016). This suggests that organic farming does possess certain difficulties that hold farmers back from taking it on a larger scale. Thus, it is necessary to identify such factors for understanding the underlying issues which could contribute in policy implication or stimulate necessary action by various stakeholders leading to the growth in the adoption rate of organic farming. With this objective, this chapter assesses factors that led some farmers to adopt organic farming system while some to practice conventional farming system.

There have been a number of organic farming-related studies in Chitwan District of Nepal where indiscriminate use of agrochemicals very much prevails (SECARD 2011) but group conversion to organic farming also exists. Adhikari (2009, 2011) finds that organic carrot and rice production system, respectively, results in higher benefit-cost ratio. A study by Bhat and Ghimire (2008) has focused on controlling major diseases and enhancing production of organic vegetables, implying scope of using bio-pesticides. Another study by Organiconepal (2006) focuses on making successful marketing method of organic products and the importance of farmers' cooperative. Only Kafle (2011a, b) has captured the issue of socioeconomic variables differing among adopters and non-adopters of organic farming system in Phoolbari VDC. In any adoption studies of agricultural innovations, socioeconomic variables are considered as important as agroecological variables and farmers' perception (Kafle 2010). This study incorporates additional variables from what were used in the previous study and expands the territorial horizon by including two more adjoining VDCs to do inclusive analysis of factors impacting the adoption of organic farming.

4.2 Socioeconomic Variables' Relation to Adoption of Farming System

Farmers' socioeconomic variables have a major role to play in farm-related decision-making, and therefore its implication on adoption of organic farming is also discussed in various studies. More so, organic farming adoption in relation to socioeconomic variables is highly context specific. Among others are HHH's age and education; the relation of which resulting in adoption of organic farming varied according to different studies. For instance, Adesope et al. (2011) assume that those who have been farming for a very long time are usually old, are less educated, and thus are more reluctant to change to organic farming. Contrastingly another study shows that older farmers with larger farms, for better-privileged relationship with extension services, are more likely to adopt organic method. They also tend to be more experienced in farming and are better educated (Alexopoulosa et al. 2010). Thus, it is also expected that with higher experience, farmers are expected to improve their expertise in organic farming. Again Khaledi et al. (2011) suggested that educated and younger farmers allocate lesser share of their cultivated area to organic farming and those with older age allocate higher share. This study also takes primary occupation of HHH as one of the indicators resulting in adoption of organic farming because it is believed that farming decision may vary with the extent of its contribution to one's livelihood. It is assumed that farming as primary occupation is expected to have positive impact on adoption as farmers would be concerned about practicing it in a more sustainable way for a long-term benefit. Likewise, those who are renting farmland are expected to have negative impact on adoption of organic farming because they would be least concerned about conserving its soil fertility for long-term sustainability.

Khaledi et al. (2011) also opined that increase in farming area would result in higher chances of not following complete adoption of organic farming because of higher labor demand. It furthermore limits complete adoption of organic farming when farmer's wage increases. Another reason could be economies of scale that can be achieved more effectively in larger conventional farms than smaller ones, and therefore for financial gain, farmers are less likely to consider a switch to organic

4.2 Socioeconomic Variables' Relation to Adoption of Farming System

farming. Contrastingly, Kafle (2011b) found farmers with large farm size to be better adopters than small farmers, probably because it signifies being resource rich and thus suggested that organic production first be promoted to large-scale farmers followed by small-scale farmers. But labor is probably one of the major defining factors among others as organic farming is labor intensive and family members have been the major source of labor in all agricultural methods irrespective of the fact that there has been increasing role of hired labor in farming practices (Pattanapant and Shivakoti 2009). Like labor, livestock holding is also an important component of organic farming as it relies mainly on manure for soil fertilization. Thus, higher livestock holding is expected to result in higher propensity to adopt organic farming. Non-farm income and social network relating to adoption of organic farming could also be observed in various literatures. Since organic farming is usually riskier in terms of production loss during initial years of conversion (Halberg et al. 2006), farmers with no source of income other than farming, which otherwise might have worked as a safety net, could feel hesitant to convert as they tend to be more risk averse. Social network is another important component that leads to participation in community activities which could provide benefit to farmers, specifically in the form of labor exchange, information sharing and knowledge gain on production, marketing, and even possibility of getting funds (Pattanapant and Shivakoti 2009; Sarker et al. 2009). Such activities could in turn make farmers participate in training and can impact to what extent farmers adopt organic farming (Kafle 2011b).

Based on field observation, group formation in Phoolbari VDC is the oldest and has conducted more training and thus is expected to have more organic farmers compared to the other two VDCs (Appendix I). Besides, there might be other unobserved factors resulting in higher adoption rate of organic farming among VDCs. Like training, access to relevant institutions like agrovets and market is expected to provide farming-related information and access to pre- and postproduction services, although its impact on adoption of organic farming could be positive or negative. For example, agrovets offer both chemical fertilizers and pesticides and packaged organic fertilizers and bio-pesticides. It depends on farmers the kind of service they desire or get influenced by as a result of easy access to it. Likewise, if there is premium market for organic products, farmers would be encouraged to practice organic farming if they are closer to the market, but in the absence of it, the case would be otherwise. Female-headed households are expected to have higher adoption rate of organic farming because they are presumed to be more health conscious for their family members. Credit is expected to have positive impact as it can provide with necessary financial accessibility for the adoption of organic farming.

Market information such as knowing the final price of a product at which a consumer buys could either encourage or discourage organic farming. Farmers would be encouraged to adopt organic farming if they know that consumers pay higher price for organic products and vice versa. One of the reasons farmer practice conventional farming is for higher profit. Thus, it is expected that the higher the extent of commercialization, the less will be the tendency to convert to organic farming. Based on the above description, the expected direction of each variable against dependent variables is presented in Table 4.1.

Table 4.1 Definition, measurement, and hypothesized relation of socioeconomic variables to adoption of farming system

Variables	Definition and measurement	Expected sign	References
HHHgender	Male-headed HH; 1 = yes, 0 otherwise	-ve	Own assumption
HHHage	Age of HHH; in years	+ve/-ve	Adesope et al. (2011), Alexopoulosa et al. (2010), and Khaledi et al. (2011)
HHHedu	Education of HHH; in years	+ve/-ve	Adesope et al. (2011), Alexopoulosa et al. (2010), and Khaledi et al. (2011)
HHHprimary_occu	Primary occupation of HHH; 1 = farming, 0 otherwise	+ve	Own assumption
Rent	Farmers renting farm land; 1 = yes, 0 otherwise	-ve	Own assumption
Org_exp	Experience of practicing organic farming; in years	+ve	Alexopoulosa et al. (2010)
LFU	Labor force availability in HH; in labor force unit (LFU)	+ve	Pattanapant and Shivakoti (2009)
LSU	Livestock holding in HH; in livestock unit (LSU)	+ve	Own assumption
Farm_size	Operational farm size; in ha	+ve/-ve	Alexopoulosa et al. (2010), Khaledi et al. (2011), and Kafle (2011b)
Non-farm_income	Income from non-farm activities (service, business, rent, remittance, pension, and laboring); in NRs./HH/year	+ve	Halberg et al. (2006)
Membership	Being in/formal group member formed for organic farming; 1 = yes, 0 otherwise	+ve	Pattanapant and Shivakoti (2009) and Sarker et al. (2009)
Org_training	Organic farming-related training; number of times	+ve	Kafle (2011b)
VDC	Belonging to Phoolbari VDC; 1 = yes, 0 otherwise	+ve	Own assumption
Agrovet	Store selling agro-products (seeds, fertilizers, pesticides, equipment, veterinary medicine, etc.); distance to nearest agrovet (in km)	+ve/-ve	Own assumption
Market	Market to selling agro-products; distance to nearest market (in km)	+ve/-ve	Own assumption
Credit	Credit taken for farming-related activities; 1 = yes, 0 otherwise	+ve	Own assumption

(continued)

Table 4.1 (continued)

Variables	Definition and measurement	Expected sign	References
Final_price	Know price of one or more crops paid by consumers; 1 = yes, 0 otherwise	+ve/-ve	Own assumption
Commercialization	Commercialization rate (total quantity of crops sold/total produced)	-ve	Own assumption

4.3 Empirical Model

The collected data is analyzed using binary logistic model (BLM). Since this study uses two different farming categories, organic farming and conventional farming, this model is applicable to assess to what extent farmers adopting one of these farming systems differ in terms of their socioeconomic variables. This model is also preferred over probit model when choice variables are mutually exclusive with each other (Long and Freese 2006).

The empirical specification for BLM can be given by:

$$E[y|x] = F(\beta'x) \tag{4.1}$$

where y is a choice variable, x is vector of explanatory variables, β is a vector of parameter estimates, and F is an assumed cumulative distribution function. Assuming F as the logistic distribution (Λ) produces the logit model, where $\Lambda(\beta'x) = \frac{\exp(\beta'x)}{(1+\exp(\beta'x))}$. Since BLM is estimated through maximum likelihood method, the coefficients cannot be interpreted as an average response of independent variables on probability of adopting organic farming. Thus, to calculate the actual magnitude of change, marginal effect is calculated which is obtained by exponentiation of the coefficient that is commonly interpreted as odds ratio (Sheikh et al. 2003).

Marginal effect in expected probability $\partial E[y]/\partial x$) is given by:

$$\partial E[y|x]\partial x = f(\beta'x)/\beta \tag{4.2}$$

where f is the corresponding probability density function (Anderson and Newell 2003).

The empirical specification for the best fit model, generated after removing redundant variables (with insignificant P-value) through backward elimination method, can be given by:

$$y = \beta_0 + \beta_1 \text{ ageHHH} + \beta_2 \text{ HHHprimary\_occu} + \beta_3 \text{ LFU} + \beta_4 \text{ LSU} \\ + \beta_5 \ln\_\text{non} - \text{farm\_income} + \beta_6 \text{ membership} + \beta_7 \text{ org}_{\text{training}} \\ + \beta_8 \text{ VDC} + \beta_9 \text{ agrovet} + \beta_{10} \text{ market} + \beta_{11} \text{ commercialization} + \varepsilon \tag{4.3}$$

where ln is natural log and ε is error term.

As per the regression rule, diagnostic tests were carried out to check the problem of multicollinearity and heteroskedasticity in the data. Though according to Pindyck and Rubinfeld (1981), variation inflation factor (VIF) is better than correlation matrix that fails to yield conclusive results, this study carries out both VIF and correlation matrix to check more vigorously the multicollinearity between any two variables. VIF indicates whether a predictor has a strong linear relationship with the other predictor(s). It gave a value of 1.55, which is below 10. According to Myers (1990), as cited in Field (2000), VIF value less than 10 indicates multicollinearity among the variables which do not exist. Similarly, Field (2000) also asserted that if any two variables have correlation value above 0.80 or 0.90, it means that they correlate very highly. Again in this case, no combination of two variables showed value more than 0.80, thus assisting us to conclude that there is no problem of multicollinearity in the data. Likewise, Breusch-Pagan/Cook-Weisberg showed significant P-value, thus rejecting null hypothesis of homoscedasticity. It means that there are linear forms of heteroskedasticity. White's test did not show significant P-value implying that there is no problem of nonlinear forms of heteroskedasticity, i.e., the variance of the error term is constant. To correct heteroskedasticity of any kind, following Nhemachena and Hassan (2007), model estimation was conducted using robust standard errors. Using robust standard errors, it neither changes the significance of the model nor the coefficients but gives relatively accurate P-values and is an effective way of dealing with heteroskedasticity (Wooldridge 2006).

4.4 Socioeconomic Factors Impacting Adoption of Farming System

The probability of the model chi-square (61.77) is highly significant at 1% that supports the existence of a relationship between independent and dependent variables. The pseudo R^2 suggests that 25.04% of the total variation in the values of dependent variable is explained by independent variables in this regression equation (Table 4.2).

In case of HHH's age, the result deviates from the findings by Khaledi et al. (2011) and Alexopoulosa et al. (2010) but adheres to Adesope et al. (2011) which showed negative relation of farmer's age with practicing organic farming. The findings suggest that a year increase in age of HHH has a highly significant negative impact on the probability of practicing organic farming by 0.5%, significant at 10%. It could be because with age, one's capacity to supply labor diminishes, which is very much necessary in the case of organic farming. As benefit from organic farming materializes only after few years of conversion, it could also be that older farmers are less willing to try new technologies because of their diminishing enthusiasm given that they will be retired soon in the near future, thus leaving less time to enjoy the benefit. There is about 4% higher probability of adopting organic

4.4 Socioeconomic Factors Impacting Adoption of Farming System

Table 4.2 Result from binary logistic model and marginal effect for organic farming system

Variables	Binary logistic model		Marginal effect
	Coefficient	P-value	
HHHage	−0.02	0.085*	−0.005
HHHprimary_occu	0.19	0.617	0.04
LFU	0.05	0.596	0.01
LSU	0.14	0.135	0.03
ln_non-farm_income	0.04	0.220	0.01
Membership	0.56	0.173	0.11
Org_training	0.53	0.000***	0.11
VDC	0.98	0.006***	0.11
Agrovet	−0.19	0.095*	−0.04
Market	0.03	0.563	0.01
Commercialization	−0.29	0.340	−0.06
Constant	−1.73	0.043**	

Source: Field survey (2013)
Note: ***1%, **5%, and *at 10% level of significance, Number of observations = 285
Wald χ^2 (11) = 61.77, Prob > χ^2 = 0.0000***
Log pseudo likelihood = −133.80232, Pseudo R^2 = 0.2504

farming if the primary occupation of HHH is farming because they will be more concerned about farming sustainably for supporting their livelihood in a long run.

Higher labor force would lead to higher probability of adopting organic farming. A unit increase in LFU will increase its chance of adoption by 1%. With higher livestock holding, which is among the fundamental components of organic farming as a source of compost for soil fertilizer, farmers' likelihood to take up organic farming goes up as well. A unit increase in LSU increases such likelihood by 3%. An increase of non-farm income by a percent also increases the chances of adopting organic farming by 1% because it will act as a safety net especially during the time of conversion when there is a risk of production failure. Membership in a group formed for the purpose of organic farming also increases the prospect of adopting organic farming by 11%. Being a member of such a group, farmers are provided with various learning platforms. Besides training, there is a high potential of knowledge generation and information gathering as a result of interaction among various stakeholders. These members meet on a monthly basis to update with their saving and loan activities. Moreover, they also gather on numerous other occasions that are irregular in nature such as meeting with NGOs, government officials, organic certifying inspector, or other stakeholders, for study trips, and while collectively selling organic products through a cooperative. But membership does not alone effect farmers' decision to convert as not all member farmers are practicing organic farming or are engaged in related activities with the similar level of keenness. That is why training plays a major role in adoption of organic farming.

Taking one more training will increase the probability of organic farming by 11%, which is highly significant as well. Training is provided by academicians and nongovernmental and government organizations. One of the regular trainings

Fig. 4.1 Farmers inspecting cabbage during Farmer's Field School (*Source*: Field survey 2013)

conducted is FFS in which the group usually meets on a weekly basis where they learn-by-doing by assessing one crop at a time from as early as its plantation period till the time of harvest (Fig. 4.1). Farmers usually divide groups to be in charge of growing a certain crop based on various organic means such as FYM, bio-pesticides, vermicompost, mulching, and so on. They discuss the amount of inputs required, problems related to pests and diseases and its management, and finally the amount harvested. Such learning process can take up to 16 weeks for each crop. Through such activity, farmers then try to replicate the most successful method in practice as well (Fig. 4.2).

Phoolbari VDC has the probability of having higher number of organic farmers by 11%, and it is significant at 1% as well. The group in Phoolbari VDC was established well before the groups in other two VDCs (Appendix I), and accordingly they have received more trainings related to organic farming (Table 4.3). Thus, it can also be suggested that the number of years these groups have been into existence and how vibrant they are into learning through programs such as FFS also has positive impact on possibility of more farmers practicing organic farming. Besides that, the distance to agrovet and market is farther from Phoolbari VDC than from other two VDCs. In addition to these, there might be other unobserved characteristics of Phoolbari VDC that resulted in more farmers adopting organic farming.

One more kilometer farther distance to agrovet decreases the probability of practicing organic farming by 4%, significant at 10%. Agrovets, in addition to offering chemical fertilizers and pesticides, are also selling commercially available packaged bio-fertilizers and bio-pesticides. Thus, uneasy access to such inputs

4.4 Socioeconomic Factors Impacting Adoption of Farming System

Fig. 4.2 Farmers discussing facts and problems encountered from each plot during Farmer's Field School (*Source*: Field survey 2013)

Table 4.3 Differentiating factors across village development committees

Variables	VDC (Mean ± SD)			T-test
	Phoolbari	Others	Total ($n = 285$)	
HHHage	50.55 ± 11.96	49.32 ± 11.74	49.97 ± 11.85	0.440
HHHedu	7.38 ± 5.53	6.44 ± 5.48	6.94 ± 5.52	0.201
Org_exp	5.06 ± 8.95	1.5 ± 5.82	3.38 ± 7.83	0.001***
Org_training	1.62 ± 2.23	0.73 ± 1.73	1.2 ± 2.05	0.001***
LFU	4.11 ± 1.78	4.44 ± 1.84	4.26 ± 1.81	0.174
LSU	2.12 ± 1.93	1.80 ± 1.48	1.97 ± 1.73	0.163
Farm_size	0.58 ± 0.46	0.51 ± 0.38	0.55 ± 0.42	0.167
Farm_income	12.00 ± 0.84	11.94 ± 0.74	11.97 ± 0.79	0.552
Non-farm income	8.98 ± 5.41	9.38 ± 5.06	9.17 ± 5.24	0.565
Agrovet	2.12 ± 1.66	1.47 ± 1.84	1.81 ± 1.77	0.006***
Market	3.57 ± 3.31	2.38 ± 3.45	3.01 ± 3.42	0.009***
Commercialization	0.94 ± 0.73	0.86 ± 0.57	0.90 ± 0.66	0.159

Source: Field survey (2013)
Note: ***1%, **5%, and *at 10% level of significance

might have discouraged farmers to take up organic farming. One more kilometer longer distance to market increases the probability of practicing organic farming by 1%. Organic farming boasts higher diversity which could be relied on for self-consumption and avoid buying or selling in the market which involves transaction costs. The higher the commercialization rate, the lesser will be the likelihood of practicing organic farming. Commercialization is mostly market oriented, which means that market-oriented farmers are less likely to practice organic farming. A unit increase in commercialization rate will decrease the probability of practicing organic farming by 6%.

4.5 Summary

The study uses binary logistic model to assess farmers' socioeconomic factors influencing their decision of adopting organic farming system and marginal effect to analyze to what extent these factors impact their decision. From this chapter, it can be recommended that while introducing organic farming, older farmers should not be prioritized for adoption of organic farming as their capacity to supply labor diminishes which is incompatible for this labor-intensive farming system. Additionally, benefit from organic farming materializes only after few years of conversion, thus diminishing their enthusiasm, as they will be retired soon in the near future which leaves them with less time to enjoy the benefit. Establishment of a group for the purpose of organic farming creates a foundation for practicing organic farming, but it is training that ultimately plays crucial role in knowledge generation and information dissemination and hence causes higher adoption rate among farmers over time. Compared to other two VDCs, farmers in Phoolbari VDC have higher probability of taking up organic farming because of group formation over a longer period of time, thus having longer experience of practicing organic farming, higher number of training, and possibly other unobserved characteristics. This study shows that farmers in other VDCs need more support and attention in their effort to practice organic farming. Agrovets also sell packaged organic fertilizers and bio-pesticides which could be the reason why farther distance to it results in lesser chance of practicing organic farming, indicating the importance of commercially available organic inputs for the vitality of organic farming.

References

Adesope O, Matthews-Njoku E, Oguzor N, Ugwuja V (2011) Effect of socio-economic characteristics of farmers on their adoption of organic farming practices. In: Sharma P, Abrol V (eds) Crop production technologies. InTech, Rijeka, pp 211–220
Adhikari RK (2009) Economics of organic vs. inorganic carrot production in Nepal. J Agricult Environ 10:23–38
Adhikari RK (2011) Economics of organic rice production. J Agricult Environ 12:97–103
Alexopoulosa G, Koutsouris A, Tzouramani I (2010) Should I stay or should I go? Factors affecting farmers' decision to convert to organic farming or to abandon it. 9th European IFSA symposium, 4–7 July 2010, Vienna
Anderson S, Newell R (2003) Simplified marginal effects in discrete choice models. Resources for the Future, Washington, DC

References

Bhat BR, Ghimire R (2008) Promotion of organic vegetable production through farmers' field school in Chitwan, Nepal. 16th IFOAM Organic World Congress, International Federation of Organic Agriculture Movement (IFOAM), Modena

Field A (2000) Discovering statistics: using SPSS for windows. SAGE, London

Halberg N, Alroe HF, Knudsen MT, Kristensen ES (2006) Global development of organic agriculture: challenges and prospects. CABI Publishing, Oxfordshire

Kafle B (2010) Determinants of adoption of improved maize varieties in developing countries: a review. Inter Res J Appl Basic Sci 1(1):1–7

Kafle B (2011a) Diffusion of uncertified organic vegetable farming among small farmers in Chitwan District, Nepal: a case of Phoolbari Village. Int J Agricult Res Rev 1(4):157–163

Kafle B (2011b) Factors affecting adoption of organic vegetable farming in Chitwan District, Nepal. World J Agricult Sci 7(5):604–606

Khaledi M, Weseen S, Sawyer E, Ferguson S, Gray R (2011) Factors influencing partial and complete adoption of organic farming practices in Saskatchewan, Canada. Can J Agric Econ 58 (1):37–56

Long ST, Freese J (2006) Regression model for categorical dependent variables using stata. A Stata Press Publication, Collage Station

Nhemachena C, Hassan R (2007) Micro-level analysis of farmers' adaptation to climate change in Southern Africa. International Food Policy Research Institute (IFPRI), Washington, DC

Organiconepal (2006) Linking organic farmers in incentive sharing mechanisms through promoting local marketing systems: An initiative from Himalayan kingdom of Nepal. Organic Agriculture Research and Production Co. (Organiconepal) Pvt. Lld., Chitwan

Pattanapant A, Shivakoti GP (2009) Opportunities and constraints of organic agriculture in Chiang Mai Province, Thailand. Asia-Pac Develop J 16(1):115–147

Pindyck RS, Rubinfield D (1981) Econometric models and economic forecasts. McGraw Hill, New York

Sarker MA, Itohara Y, Hoque M (2009) Determinants of adoption decisions: the case of organic farming (OF) in Bangladesh. Extens Farming Syst J 5(2):39–46

SECARD (2011) Market oriented organic agriculture promotion project (MOAP) in Chitwan District of Nepal. Society for Environment Conservation and Agricultural Research and Development (SECARD) Nepal, Kathmandu

Sheikh AD, Rehman T, Yates CM (2003) Logit models for identifying the factors that influence the uptake of new 'no-tillage' technologies by farmers in the rice-wheat and the cotton-wheat farming systems of Pakistan's Punjab. Agric Syst 75(1):79–95

Willer H, Lernoud J (2016) The world of organic agriculture: statistics and emerging trends. International Federation of Organic Agriculture Movements (IFOAM)/Research Institute of Organic Agriculture (FiBL), Bonn/Frick

Willer H, Yussefi M (2006) The world of organic agriculture: statistics and emerging trends. International Federation of Organic Agriculture Movements (IFOAM)/Research Institute of Organic Agriculture (FiBL), Bonn/Frick

Wooldridge JM (2006) Introductory econometrics: a modern approach. Thomson South Western, Ohio

Chapter 5
Soil Properties of Organic and Conventional Farming Systems

Abstract Soil has a dominant role to play in improving crop productivity. Farmers use both organic and conventional means of soil amendments to maintain its fertility. This chapter analyzes the difference in soil quality of organic and conventional farms that are known to rely on sustainable and unsustainable means of soil fertility management practices, respectively. The study was conducted in Chitwan District of Nepal where both kinds of farming systems are known to coexist. The soil was tested on the basis of most common parameters: texture, pH value, organic matter, and nitrogen, phosphorus, and potassium level. No significant difference was found between the two soil types except for phosphorus and potassium level, which are significantly higher in organic soil but higher overall according to the nationally defined limit. The overall soil texture is sandy loam, moderately acidic with medium level of organic matter and nitrogen. While in present context, soil quality might not deviate much between organic and conventional soil, the same cannot be implied for the future as environmental change takes place after many years of accumulated processes.

5.1 Introduction

Soil has a dominant role to play in improving crop productivity. It helps to infiltrate water and air, allows plant roots to expand, and encourages biota to thrive that in turn controls physical degradation of soil and cycles nutrients based on plant needs (Green and Brye 2008; DuPont 2012). While it has affected our ability to cultivate crops and advance civilization that we know of today, intensive cultivation and harvesting of crops for human or animal consumption has altered its natural nutrient cycling processes and devoid it of plant nutrients. To improve soil fertility, humans traditionally used animal manure, charcoal, ash, and lime as soil amendments. Today, however, farmers rely on numerous sources, including inorganic chemical fertilizers with or without supplementing organic sources of nutrients, such as manure or compost. Although excess usage of such soil amendments does not always mean optimized utilization. For example, excess nutrients, especially nitrogen and phosphorus, can result in surface runoff or leaching from agricultural fields which can pollute surface and groundwater.

In order for soil to provide its services in the best possible way, one has to make sure that it has all the ingredients available at the right amount to maintain its quality which allows for water retention and drainage, oxygen in the root zone, and nutrients to facilitate crop growth and provide physical support for plants to grow in a healthy way (Parikh and James 2012). Soil testing is one way to determine this. This chapter compares soil properties of organic and conventional farm because these two farming systems have their own method of soil fertility management practices. Organic means are known to enhance or sustain the overall quality and health of the soil ecosystem, while conventional means such as chemical fertilizer deplete soil fertility by reducing soil structure and soil aggregation, decreasing water infiltration, and increasing soil bulk density, soil salinity, nitrogen leaching, and groundwater contamination (Ikemura and Shukla 2009). In this chapter, comparison is basically made on soil texture, pH, organic matter, and three main mineral components, nitrogen, phosphorus, and potassium, which are common parameters constituting the healthy soil.

5.2 Soil Properties

Soil texture, pH, organic matter, and nitrogen, phosphorus, and potassium level are the common parameters for soil testing. Soil texture indicates the relative proportion of particles of various sizes such as sand, silt, and clay. Sandy soil feels gritty, silty soil feels smooth like talcum powder, whereas clay soil feels harsh when dry and slippery and sticky when wet. Sand particles constitute the biggest pieces of soil particles (Vinje 2016) and are generally round with largest in size among all (0.05–2 mm), followed by silt (0.002–0.05 mm) and clay (0.002 mm) that are usually thinner and flatter. Therefore, higher sand (larger, round particles) content in soil means more space available for water and air to pass through for plants. But sometimes air spaces in sandy soil are too large to hold water against the gravity, causing soil to have lower water holding capacity and thus making it prone to drought (DuPont 2012). Similarly, air in the soil converts atmospheric nitrogen into a usable form for plants, but excessive amount leads to rapid decomposition of organic matter. It also causes nutrient loss and tends to have low level of beneficial microbes and organic matter that plants thrive on. On the other hand, soils with heavy clay are quite dense, difficult to drain, hard, and crack when dry. Because of lack of space between the clay particles, soil gets waterlogged, suffocating plant roots and soil organisms. Organic matter or microbial life cannot thrive much in such soil, and plant roots cannot expand in such hard material. Silty soils are also dense, but it is considered to be more fertile than either sandy or clay soils. In any case, the best soil should consist both small and large pore spaces (Vinje 2016). Soil texture is inherent and usually difficult to change. On the other hand, soil structure, which is the arrangement of soil particles, can be manipulated by various management practices of plowing, cultivating, adding lime or organic matter, and stimulating biological activity to improve soil quality (DuPont 2012).

5.2 Soil Properties

Soil pH reading shows acid-alkaline balance in soil and is an important chemical property as it controls a wide range of physical, chemical, and biological processes and properties that affect soil fertility and plant growth. It significantly influences availability of nutrients to plants, activity of microorganisms in the soil, and even stability of soil aggregates (Rosen et al. 2016; Parikh and James 2012). Organic matter is a partial or well-decomposed residue of organic biomass available in soil that provides essential plant nutrients, beneficially influences soil structure, buffers soil pH, and improves water holding capacity and aeration (Parikh and James 2012). Organic matter content in soil holds soil particles together in aggregates to improve soil structure and help maintain moisture. They absorb and store nutrient for plants to use when they require and are a source of food for beneficial microorganisms (DuPont 2012). It also balances out the soil, whether acidic or alkaline (Vinje 2016).

Optimum crop production depends on at least 17 essential elements required for plant growth: carbon (C), hydrogen (H), oxygen (O), nitrogen (N), phosphorus (P), potassium (K), calcium (Ca), magnesium (Mg), sulfur (S), iron (Fe), manganese (Mn), zinc (Zn), copper (Cu), boron (B), molybdenum (Mo), chlorine (Cl), and nickel (Ni). Among these, plants obtain carbon, hydrogen, and oxygen from air and water, while the remaining components are derived from the soil (Rosen et al. 2016). Of the 17 essential elements for plant growth, N, P, and K are the most important ones and are known as primary macronutrients as plants require them from soil in larger amounts (Vinje 2016). Among these too, N is the most vital component as it is required by plants in the largest quantity but has limited availability during the growth phase because it is easily lost from the soil system. It is an essential element of all amino acids that are the building blocks of protein. It is a component of nucleic acid such as DNA that holds genetic code to grow and reproduce all living things. It is also a major component of chlorophyll that allows for the process of photosynthesis (CTAHR 2016). Phosphorus is needed for storage and transfer of energy in the plant. It is essential in every metabolic process, protein synthesis, sugar development, and legume nitrogen fixation. It is crucial for root development, rapid seedling growth, winter hardiness, disease resistance, efficient water use, early maturity, and maximum yield. Potassium is a regulator of metabolic activities and is essential for photosynthesis and protein synthesis as well as carbohydrate transport and storage. It promotes root reserves, winter hardiness, cell development, and strong walls and reduces stalk lodging. It also improves water use efficiency, increases yield, improves crop quality, and reduces incidence of disease (Rosato 2016). In general, these three macronutrients are most likely to be in short supply in agricultural soil than required by the plants. Calcium, magnesium, and sulfur are categorized as secondary macronutrients that are needed in smaller quantities and are usually available in sufficient quantities in the soil. The remaining components called micronutrients, also known as trace nutrients, are needed in much smaller amounts (Vinje 2016; Parikh and James 2012).

5.3 Methodology of Soil Testing

Chitwan District in southern Nepal was selected as a study site because the use of agrochemicals is quite common (SECARD 2011) but at the same time group conversion to organic farming also exists which makes it possible to get soil sample of both farming systems from the same area. It lies in the plain area which has a climate of subtropical monsoon with an average annual rainfall of 2318 mm, thus endowing it with high agricultural potential (Devkota et al. 2011). Due consideration should be given in the process of taking soil sample in order to legitimize the interpretation of soil test results. Starting with the time of the year, although there is no such barrier, spring and fall sampling are usually considered to be better when soil is most stable (Rosen et al. 2016; Vinje 2016). The sample was taken in March 2013, which is the beginning of spring in Nepal. Second step is to mark random areas within a farm from where the subsamples are to be collected. This is to integrate soil texture of different levels as a result of previously grown crops and application of inputs such as fertilizer, organic amendments, lime, etc. About 5–10 subsamples for relatively small areas (less than 1000 square feet) are recommended (Rosen et al. 2016), which was duly followed. It is important to collect samples from appropriate depths that actually matters for plant growth. Usually the tillage depth in 6-inch intervals is considered to be appropriate for sampling depth. The most root activity and fertilizer applications are generally restricted to the top 6 inches of soil and are typically used for conventional tests of organic matter, P, K, pH, and salt levels (USDA n.d.). Thus, all the surface vegetation or litter were scraped off to take subsamples from surface till the depth of about 6 in. by digging a V-shaped pit using a spade. All the subsamples were then mixed thoroughly, and not more than 0.25 kg per field was taken as a final sample that was then kept in a plastic bag and labeled properly. The total of 30 samples was taken, 15 for organic farm and another 15 for conventional farm. Samples were then sent to Soil Testing and Service Section of Department of Agriculture, which is the national soil testing facility in Nepal. All the variables are analyzed using t-test to see if there is any significant difference among organic and conventional soils. Soil texture triangle (STT) (Fig. 5.1) was used to know the soil type based on the result of proportion of soil particles: sand, silt, and clay. STT indicates classes which break the distribution of particle sizes or soil textures into 12 categories: clay, sandy clay, silty clay, sandy clay loam, clay loam, silty clay loam, sand, loamy sand, sandy loam, loam, silt loam, and silt (USDA 1987). The values of each variable are interpreted based on definition of soil fertility given by Nepal Agricultural Research Council (NARC 1993) as cited in Maharjan (2010), which is provided in Table 5.1.

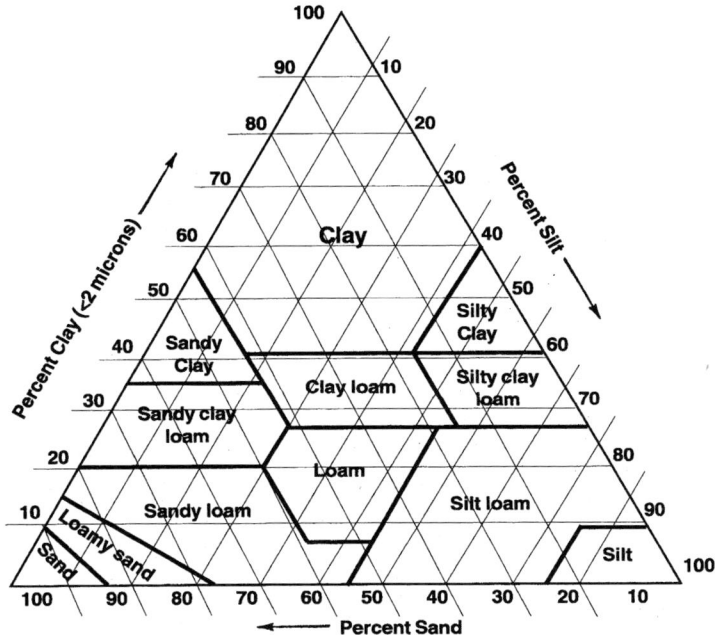

Fig. 5.1 Soil texture triangle (*Source*: Sandall (2016))

5.4 Difference in Properties of Organic and Conventional Soil

Table 5.2 shows result of t-test analysis among various components of soil in organic and conventional farm. The result shows that there is no statistically significant difference in the means of all the variables measured except for P and K which are significantly different at 5% and 10% level, respectively. Based on STT, the texture of the soil is sandy loam, which means it has good water infiltration, aeration, and workability but poor nutrient and water holding capacity (YDAE 2016). Thus, plants grown in this type of soil will require regular fertilization and irrigation.

Soil pH ranges from the scale of 0 to 14, with lower range indicating acidity, 7 indicating neutrality, and higher range indicating alkaline. In this case, the overall average value of soil pH is 5.39, which means that soils tested are acidic. According to NARC, this value of pH is moderately acidic. Many plants including commonly grown fruits, vegetables, flowers, trees, and shrubs can tolerate pH value ranging between 5.2 and 7.8 (Rosen et al. 2016). Soil pH between 5.5 and 8.5 is optimal for compost microorganisms, while range of 6–7.8 is needed for nutrients required for plant growth to be water soluble in soil. Beyond this, micronutrients and P become less available to roots. Below 5.5 pH, many of the major nutrients become less available, and some of the micronutrients can become toxic to the roots system

Table 5.1 Interpretation of soil parameters

pH value		Organic matter (%)		Total nitrogen (%)		Available phosphorus (kg/ha)		Available potassium (kg/ha)	
Range	Level	Range	Level	Range	Level	Range	Level	Range	Level
<4.5	Strongly acidic	<2.5	Low	<0.1	Low	<31	Low	110	Low
4.5–5.5	Moderately acidic	2.5–5	Medium	0.1–0.2	Medium	31–55	Medium	110–280	Medium
5.5–6.5	Weakly acidic	>5.0	High	>0.2	High	>55	High	>280	High
6.5–7.5	Nearly neutral	–	–	–	–	–	–	–	–
>7.5	Alkaline	–	–	–	–	–	–	–	–

Source: NARC (1993) as cited in Maharjan (2010)

5.4 Difference in Properties of Organic and Conventional Soil

Table 5.2 T-test of soil properties

Variables	Farming systems (Mean ± SD)			T-test
	Organic ($n = 15$)	Conventional ($n = 15$)	Total ($n = 30$)	
Sandy (%)	62.19 ± 3.25	57.36 ± 16.91	59.77 ± 12.21	0.287
Silty (%)	28.73 ± 3.62	29.84 ± 6.14	29.29 ± 4.99	0.553
Clay (%)	8.77 ± 0.76	8.8 ± 0	8.79 ± –	0.893
pH	5.40 ± 0.36	5.37 ± 0.47	5.39 ± 0.41	0.843
Organic matter (%)	2.63 ± 0.33	2.73 ± 0.53	2.68 ± 0.44	0.521
Nitrogen (%)	0.12 ± 0.01	0.12 ± 0.02	0.12 ± 0.01	0.526
Phosphorus (kg/ha)	350.36 ± 208.35	196.14 ± 177.75	273.25 ± 205.82	0.038**
Potassium (kg/ha)	534.66 ± 341.18	324.91 ± 275.33	429.78 ± 322.75	0.075*

Source: Field survey (2013)
Note: **5% and *at 10% level of significance

(A&L Canada Laboratories 2004). Low pH level also does not allow essential plant macronutrients (i.e., N, P, K, Ca, Mg, and S) to be bioavailable than higher pH value near 7. With pH range 5.0 to 6.0, certain micronutrients (i.e., Fe, Mn, Zn) tend to become more soluble and potentially toxic to plants. At low pH value less than 5.5, aluminum toxicity tends to be a common problem for crop growth. Soil pH values from 6 to 7.5 are considered to be optimal for plant growth, although certain plant species grow better in more acidic or basic conditions.

To raise soil pH, amendments such as lime can be used, while ammonium sulfate, iron sulfate, or elemental sulfur can be added to soil to lower pH (Parikh and James 2012). Manures have high pH value of 8.0 to 8.5, while by-products of wood residuals or peat moss have pH value as low as 4.5. Since different level of pH is demanded by different plant species, it will have to be adjusted accordingly (Darlington n.d.). Adding ground limestone by about 3–4 pounds per 100 square feet in sandy soil and 7–8 pounds of limestone per 100 feet for loamy soil is recommended to raise soil pH. It should be applied at least 2 to 3 months before planting any crops for it to work. Wood ash can also raise the pH of soil, but only light amount should be applied on soil top in the fall and thoroughly turn the soil in the spring. Too much wood ash may result in nutrients runoff from the soil, and when seed comes in contact with wood ash, it reduces the probability of seed germination. It usually takes a season or two to bring the pH value to moderate level, and afterward little effort per year would suffice to maintain it (Vinje 2016).

The average value of organic matter in the tested sample is 2.68%, which is at the medium level. Organic matter can be added in the form of compost, matured compost, and using mulch or growing cover crops as green manures. Compost can improve the texture and nutrient level of the soil. Mulching by using straw, hay, grass clippings, and shredded bark will protect the soil from extreme heat and cold, reduce water loss through evaporation, and prevent weed growth as well (Vinje 2016). Overall average N level of 0.12% suggests that it is also available at the medium level, based on NARC's definition. For N-deficient soil, blood meal, fish emulsion, alfalfa meal, or alfalfa pellets should be used. Alfalfa as cover crop can also make N available to plants. Cottonseed meal can also provide N to the soil, but due to its acidic nature, it should be used in combination with lime. Spring or fall is considered to be the best time to add any soil amendments or organic fertilizer if the soil is devoid of minerals or nutrients because it is most stable during this period (Vinje 2016).

The average phosphorus value of organic soil is 350.36 kg/ha, which is significantly higher at 5% than conventional soil with 196.14 kg/ha. The potassium value of organic soil is also significantly higher than conventional soil at 10%, with the value of 534.66 kg/ha and 324.91 kg/ha, respectively. Heinrich et al. (2017) have also reported high P and K levels on a long-term organic farm. The reason is that most of the organic fertilizers and amendments such as manures and manure-based composts and fertilizers that organic farm relies on for N supply, which they need in larger amount than any other element, contain other nutrients too. For example, chicken manure typically has an analysis of 4-3-3 (N-phosphate (P_2O_5)-potash (K_2O)). Unlike N, P and K are relatively immobile in most soils, meaning they cannot be washed off the root zone with irrigation or rain, and build up over time. In this case, the average P and K value is very high in both soils compared to the one recommended by NARC. Like organic farmers, conventional farmers too rely on organic fertilizers in addition to chemical ones. This could be the reason why there is excess P and K supply in the soil than is needed by a crop.

High phosphorus leads to sediment loss through runoff, which when enters waterways might contribute to an increase in algal growth (eutrophication), thus ultimately resulting in death of aquatic organisms. While high level of K is not harmful to the environment, it can limit the ability of a plant to uptake Ca and Mg. The ruminant livestock (beef and dairy cattle and sheep), when fed forage grown on high K soil, has higher chance of getting metabolic disease as a result of magnesium deficiency (Heinrich et al. 2017). As a precaution, bone meal and softrock phosphate should be avoided as they increase the P level (Vinje 2016), and K fertilizer should be reduced to benefit economically. Between these two elements, K is much more likely to be removed by crops in larger amounts than P from the soil during a growing season (Heckman n.d.). But since in this case, both are at the higher level, some recommendations provided by Heinrich et al. (2017) to supply N without adding P and K to the soil could be implemented. Growing forage crops helps remove P and K from the system; N-fixing winter cover crops (legumes) supply sufficient N needed by the crop at the lowest cost compared to other organic fertilizers without increasing P or K. Making compost from on-farm organic materials cycles nutrients on the farm without increasing P and K. Some vegetables (Table 5.3) can also help remove high amount of P and K when harvested. The use of N-only fertilizers or with a high N analysis relative to P and K such as feather meal that has ratio of 13-0-0 (N-P_2O_5-K_2O) fertilizers and nutrient budgeting the inputs and outputs of elements by focusing on the kind of crops to grow are other ways to control P and K in the soil. However, there are added responsibilities or challenges in practicing these options. Excess forage crops that are grown for balancing N-P-K in the soil, for instance, will have to be managed in some form such as selling to an off-farm animal operation. One requires experience and

Table 5.3 Estimates of phosphorus and potassium removal by vegetable crops

Crops	Yield (ton/acre)	P_2O_5 (lb*/acre)	K_2O (lb/acre)
Broccoli	8	20	110
Cabbage	30	55	230
Carrots	15	25	100
Cauliflower	6	20	60
Cucumber, slicing	10	10	40
Lettuce, romaine	20	30	170
Onion, bulb	34	50	160
Peas, shelled, bush	2	10	20
Peppers, bell	20	30	110
Potatoes	20	60	250
Snap beans, bush	6	15	40
Spinach	12	15	120
Squash, summer	20	30	130
Squash, winter	18	20	120
Sweet corn	10	30	60
Tomato	12	10	80

Source: Heinrich et al. (2017)
Note: *lb stands for pound

equipment to make compost from on-farm organic materials. When certain vegetables are prioritized just for their ability to take off more P and K from the soil, one needs to think of its market potential as well. Similarly, the drawback of using feather meal is that it is costlier and is available at the limited amount. And finally, nutrient budgeting might seem like a very appealing concept, but it is time-consuming to grow diverse crops and comes with uncertainty in estimating each crop's uptake and removal.

Although this study did not find any significant difference in soil quality of organic and conventional farming, it cannot be assumed that the situation would be the same over time. Moreover, the longevity in the use and quantity of chemicals also matters, which regrettably this chapter does not consider. According to Rigby and Caceres (1997), the concept of time horizon is also important to understand the concept of sustainability as usually environmental change takes place after many years of accumulated processes. Thus, agro-system that appears sustainable today could be in the process of being unsustainable over a long time period.

5.5 Summary

The comparison of organic and conventional soil showed that there is no significant difference in the soil texture, pH value, organic matter, and nitrogen level. The only difference was in phosphorus and potassium level, which are significantly higher in organic soil. Based on soil texture triangle, the overall soil texture is sandy loam, which means it has good water infiltration, aeration, and workability but poor nutrient and water holding capacity. Thus, plants grown in this type of soil will require regular fertilization and irrigation. Other variables were measured based on the desired level suggested by Nepal Agricultural Research Council. The overall average value of soil pH is 5.39, which suggests soils tested are moderately acidic. Since pH value closer to 7 is desirable for healthy plant growth, amendments such as lime, manure, and moderate amount of wood ash are recommended to raise soil pH. With an average value of 2.68%, organic matter in the tested sample is available at the medium level. Compost, manure, or mulching by using straw, hay, grass clippings, and shredded bark is suggested to raise organic matter. Overall, average nitrogen level of 0.12% shows that it is also available at the medium level, whereas phosphorus and potassium value is very high in both kinds of soils. For nitrogen-deficient soil, blood meal, fish emulsion, and alfalfa should be used. In order to avoid or lessen excess amount of phosphorus and potassium, certain measures can be taken such as growing forage crops, legumes, and certain vegetables, making compost from on-farm organic materials, using nitrogen-only fertilizers or with a high nitrogen analysis relative to phosphorus and potassium, and nutrient budgeting. However, there are added responsibilities or challenges associated with these practices. While in the present scenario, there seems to be not much difference in organic and conventional soil, the same cannot be implied for the future as environmental change takes place after many years of accumulated processes.

References

A&L Canada Laboratories (2004) Compost management: compost analysis for available nutrients and soil suitability criteria and evaluation. A&L Canada Laboratories Inc., London
CTAHR (2016) Nitrogen. College of Tropical Agriculture and Human Resources. University of Hawaii. http://www.ctahr.hawaii.edu/mauisoil/c_nutrients01.aspx. Retrieved 9 Feb 2016
Darlington W (n.d.) Compost: a guide for evaluating and using compost materials as soil amendments. Soil and Plant Laboratory, Inc., California
Devkota R, Budha PB, Gupta R (2011) Trematode cercariae infections in freshwater snails of Chitwan district, central Nepal. Himal J Sci 7(9):9–14
DuPont ST (2012) Introduction to soils: soil quality. Pennsylvania State University, Pennsylvania
Green VS, Brye KR (2008) Soil quality: an essential component of environmental sustainability. Electron J Integr Biosci 6(1):1–2
Heckman JR (n.d.) Soil fertility test interpretation: phosphorus, potassium, magnesium and calcium. Rutgers Cooperative Extension, The State University of New Jersey, New Jersey
Heinrich A, Falen J, Stone A (2017) High soil test phosphorus and potassium levels on a long-term organic farm: trends, causes and solutions. eOrganic
Ikemura Y, Shukla MK (2009) Soil quality in organic and conventional farms of New Mexico, USA. J Organ Syst 4(1):34–47
Maharjan M (2010) Soil carbon and nutrient status of rangeland in upper Mustang. Institute of Forestry, Tribhuvan University, Pokhara
Parikh SJ, James BR (2012) Soil: the foundation of agriculture. Nature Educ Knowl 3(10)

References

Rigby D, Caceres D (1997) The sustainability of agricultural systems. Institute for Development Policy and Management. University of Manchester, Manchester

Rosato C (2016) Balancing soil using organic minerals. https://woodleaffarm.com/enlivening-soil/ . Retrieved 5 Mar 2016

Rosen CJ, Bierman PM, Eliason RD (2016) Soil test interpretations and fertilizer management for lawns, turf, gardens and landscape plants. Regents of the University of Minnesota, Minnesota

Sandall L (2016) Physical properties of soils http://croptechnology.unl.edu/pages/informationmodule.php?idinformationmodule=1130447123andtopicorder=2andmaxto=73andminto=1. Retrieved 9 Feb 2016

SECARD (2011) Market oriented organic agriculture promotion project (MOAP) in Chitwan district of Nepal. Society for Environment Conservation and Agricultural Research and Development (SECARD) Nepal, Kathmandu

USDA (1987) Soil mechanics level I: module 3 USDA textural soil classification. United States Department of Agriculture, Washington, DC

USDA (n.d.) Sampling soils for nutrient management. United States Department of Agriculture (USDA), Washington, DC

Vinje E (2016) Planet Natural. www.planetnatural.com/garden-soil/. Retrieved 29 Aug 2016

YDAE (2016) Soil texture. http://www.ydae.purdue.edu/natural_resources/Soil,Health/Activities/SoilTexture2,SWS2.pdf. Retrieved 2 Sept 2016

Chapter 6
Crop Management Through Organic Means

Abstract Organic means of crop management practices in this chapter indicates soil and pest management practices. It is an imperative issue to overcome challenges of declining soil fertility along with the need to increase food production for growing population. The main objective of this chapter is to assess how adoption of organic means of crop management practices differs among organic and conventional farmers with various socioeconomic characteristics, as in developing countries, even though there is an influx of modern inputs such as chemical fertilizers, farmers still incorporate traditional ways of soil management practices such as farm yard manure. Clearly there is lack of understanding on the extent of adoption of such options in order to guide in developing farm-level adoption strategies. The collected data was analyzed using multivariate probit model which regressed five soil fertility management practices, viz., mulching, compost-shed, bio-slurry, vermicomposting, and plastic cover, in addition to bio-pesticides for pest management against various socioeconomic variables of 285 organic and conventional farmers. Results show that mulching is the most traditional form which is still practiced under both farming systems, although the rest are mostly adopted by organic farmers. Fund assistance/credit availability contributes positively to adoption of high investment requiring practices such as compost-shed and bio-slurry, while bio-pesticides can be appealing to less resource holding farmers as it can be made from locally available resources as well. One of the ways to increase adoption rate is training as it complements technical knowledge required to implement these practices. Moreover, farmers are complementing one practice with another and some even act as substitutes. Thus, any effort to enhance such adoption rate can consider these characteristics of various practices.

6.1 Introduction

Crop management practices in this chapter indicate organic means of soil and pest management practices. Soil is a principal component that influences farming productivity. It is the basis for plant growth by supplying nutrients, water, and root support. It maintains biodiversity by providing habitat for billions of organisms. Soil fertility is a result of both inherent and dynamic soil properties. Inherent soil property is a result of natural soil-forming processes, whereas dynamic soil property is defined by how well it is managed by humans (Green and Brye 2008). Therefore, adopting various management practices can impact on dynamic properties of soil. Soil management practices deserve greater attention for various reasons, one among which is the issue of food insecurity. It is one of the major challenges facing the world today and is more prevalent in developing countries where agriculture remains a major sector. Declining soil fertility to the large extent is responsible for lower productivity, and hence better soil management practice could be one of the ways to combat this situation. The fact that there is continuous growth in demand for producing more on a limited area requires us to focus on management strategies that lead to better soil fertility for enhanced productivity (Green and Brye 2008; Huili et al. 2013; OECD 2008). Nepal remains no exception and faces similar challenges of declining soil fertility along with the need to increase its food production for growing population. Declining soil fertility is a result of changes in agricultural practices through changes in technology and farmers' knowledge. Soil erosion, organic matter depletion, acidification, degradation of forest and marginal land, crop intensification, and insufficient and unbalanced use of chemical fertilizer are the major reasons for soil fertility depletion in Nepal (Bista et al. 2010; Shrestha et al. 2013).

Usually organic farmers are known for being better managers of soil. It relies on managing soil organic matter that enhances chemical, biological, and physical properties of soil, thus optimizing the crop production. The difference between organic and conventional way of soil management is that the former relies on longer-term solutions with the objective of preventing rather than reacting and the latter is based on short-term solutions (Watson et al. 2002). However, in developing countries, even though there is an influx of modern inputs such as chemical fertilizers, farmers still incorporate traditional ways of soil management practices such as FYM (Bista et al. 2010; Kabuli and Phiri 2007). As for managing the pests, bio-pesticides are known to be the nontoxic alternative to the conventional means (EPA 2013) which means that unlike chemical pesticides, bio-pesticides do not leave residue in food which is detrimental for our health. The previous study by Bhat and Ghimire (2008) on bio-pesticides in the study areas only deals with its scope of usability in controlling major diseases and enhancing production of organic vegetables, but does not analyze adoption rate of such technologies which is much more important as it reveals the benefits or problems encountered in actually practicing it. Therefore, the main objective of this chapter is to assess how adoption of organic means of crop management practices (OCMPs) differs among organic and conventional farmers with various socioeconomic characteristics. Understanding the determinants of farmers' choices of OCMPs among the various available choices can provide insight on the factors that enable or constrain such actions and guide in developing farm-level adoption strategies.

6.2 Socioeconomic Variables' Relation to Adoption of Organic Means of Crop Management Practices

This chapter considers five organic means of crop management practices: mulching, compost-shed, bio-pesticides, bio-slurry, and others (vermicompost and/or plastic cover). Mulching is a process of covering soil surface around the plants to conserve moisture content, protect plant roots, reduce weed growth, and improve overall soil health and fertility. Farmers in the study area are practicing either crop or plastic mulch, which are then combined to form a variable "mulch." In study areas, crop mulching is done through the use of hay, straw, forage, or other cover crops. Comparatively, mulching through crop residues is more traditional and cheaper, and in addition to that, it also supplies nutrients (Schonbeck 2012). It was found that farmers are either applying FYM or compost for soil management. FYM is simply an animal excrement mixed with hay, leaves, and any other decomposable material from the field, while compost is the state when such mixture is allowed to be decomposed. According to OEFFA (n.d.), compost is defined as "the product of a managed process through which microorganisms break down plant and animal materials into more available forms suitable for application to the soil," whereas manure is defined as "feces, urine, other excrement, and bedding produced by livestock that has not been composted." This study takes compost-shed (Fig. 6.1) as a proxy for quality FYM or compost because it preserves the compost/manure pile from volatilization by sun or leaching by rainfall and maintains its nutrient availability (Fig. 6.2) (Bista et al. 2010).

The survey found that farmers are either preparing bio-pesticides by themselves (Fig. 6.3) or are buying from the market (Fig. 6.4). Bio-pesticides are meant for controlling pests through nontoxic means and thus have no harmful effect on soil unlike the conventional pesticides (EPA 2013). Bio-slurry, on the other hand, is a by-product obtained from biogas plant after the dung or other biomasses have been digested for the generation of gas. It also revitalizes soil fertility, and this study takes biogas as a proxy for farmers applying bio-slurry. The variable "others" include those using plastic cover and/or vermicompost. Plastic cover is the way of covering the plants with a plastic-clad semicircular structure. It helps increase production and improve quality by managing soil moisture and subsequently making nutrients available (Montri and Biernbaum 2009) which means that farmers will not have to or rely less on chemical inputs that has detrimental impact on soil over time. Vermicompost is yet another high-quality compost produced from worm castings. Since the number of respondents undertaking these (whether separately or combined) was very limited (only 6% of farmers used plastic cover (Fig. 6.5) and 7% used vermicompost), it has been combined to form the variable "others."

Literatures were reviewed to learn how adoption of soil conservation or sustainable pest management technologies might differ among farmers with various socioeconomic variables. Most of the studies were conducted on adoption of integrated soil fertility management (ISFM), which comprises of mineral fertilizers, locally available soil amendments (such as lime and phosphate rock) and organic

Fig. 6.1 Improved compost-shed (*Source*: Field survey 2013)

Fig. 6.2 Farm yard manure that is kept in open, risking volatilization by sun or leaching by rainfall (*Source*: Field survey 2013)

matter (crop residues, compost, and green manure). ISFM emphasizes on locally acceptable practices leading to nutrient and water use efficiency, thus increasing the agricultural productivity (IFDC 2014). The soil management practices incorporated in this study also aim to improve soil fertility for enhanced production. Hence, it could be assumed that the socioeconomic variables will have similar impact on

Fig. 6.3 Self-made bio-pesticide prepared in a plastic bin using local resources (*Source*: Field survey 2013)

Fig. 6.4 Bio-pesticide available in the market (*Source*: Field survey 2013)

adoption of organic means of pest management practices because it also relates with sustainable practices.

Studies show that different socioeconomic variables will have varying impact on adoption of new technologies. The age of HHH's influence on a decision to adopt may be positive or negative. The older farmers might feel reluctant to change their old ways compared to younger ones who are more knowledgeable about new practices and who might want to take risk of trying out new technology because of their farsighted vision in farming. But on the other hand, older farmers have more

Fig. 6.5 Crop cultivation using plastic cover (*Source*: Field survey 2013)

experience, resources, and authority, which might induce them to adopt new technologies as well (Akinola and Owombo 2012). Study by Mugwe et al. (2009) agrees with the former reasoning of impact of age on technology adoption. Studies basically relate age with experience of farmers and conceive to have similar impact on technology adoption. It is revealed that experienced farmers are less likely to adopt new technologies because less experienced ones, due to lack of knowledge, are more responsive toward new technologies and with experience, they realize that such technologies are not suitable for the local ecosystem (Kuntariningsih and Mariyono 2013). In this case, bio-pesticides and vermicompost are comparatively new practices. Another reason could be that experienced farmers are more likely to retire in the near future which means that they will have less time to reap the benefit from such investment (Grazhdani 2013). In this chapter, farming experience is taken as years of practicing organic farming. So it is expected that it could have either positive or negative impact because with experience of organic farming, farmers are anticipated to be more competent in adopting organically viable practices, but on the other hand, as shown by literatures, experienced farmers indicate age too and so they might not find it beneficial over a short time period that they have before they retire or they just realize the unfeasibility of such technology.

Education possesses ability to obtain, process, and use new information and therefore could have positive influence in the adoption of technologies that require technical knowledge (Akinola and Owombo 2012; Grazhdani 2013; Adolwa et al. 2010). Household size can also have both positive and negative influence on adoption of these technologies. It is usually taken as a proxy for labor. With more members, there will be higher labor supply to adopt and practice new technologies.

But more members also mean more pressure for consumption, meaning labor might have to be diverted for earning higher income. Thus, it is difficult to generalize how it will impact on adoption of these technologies (Akinola and Owombo 2012). However, studies by Kuntariningsih and Mariyono (2013) and Mugwe et al. (2009) agree with the former assertion, showing that higher supply of labor is an indication of technology adoption because such technologies generally require more labor. Livestock holding is shown to have inverse relation to technologies such as mulching because in mixed farming system, livestock holding and crop mulching compete for crop residue (Jaleta et al. 2012). But study by Mugwe et al. (2009) showed that less livestock holding implies less manure supply which is why farmers will be willing to look for other alternatives or will try to manage in a better way so as to maximize the effectiveness of small quantities of manure they have to improve soil fertility. Adolwa et al. (2010) suggests that livestock holding is an indication of resource endowment and are more likely to look for information on new agricultural technologies and subsequently practice it.

It is assumed that larger farm size also results in higher adoption rate because it signifies increased availability of capital, which makes investment in such technologies more feasible. It can also be implied that those with small farm size are usually risk averse and thus hesitate to invest in technologies with uncertain results in their limited farm (Akinola and Owombo 2012; Grazhdani 2013). In a study by Grazhdani (2013), farm income is a critical variable showing positive correlation with adoption of conservation-oriented farming. Technology being a normal good, it is perceived that farmers with an intention of amplifying the farm income will be more inclined into implementing these technologies. Non-farm income provides the much needed supply of capital, which makes adoption of these technologies feasible (Akinola and Owombo 2012). Contrastingly, Adolwa et al. (2010) found non-farm income to negatively influence the adoption of soil management technologies, as people preferred investing such income in non-farm activities. Training is undoubtedly an important component, which encourages farmers to adopt these technologies. However it may not have any significant impact if farmers already are knowledgeable about the importance but lacks resource to purchase and implement such technologies (Bizimana 2013).

Referring to reviewed literatures, Table 6.1 provides the hypothesized relation of various socioeconomic variables of farmers against the OCMPs, which they have adopted. As for the farming system, it is expected that organic farmers will adopt all of the practices incorporated in this study since they will be more conscious of improving soil fertility and managing pests through organic means. Conversely, conventional farmers are expected not to adopt any of these practices as it is assumed that they will prefer to rely on conventional means for such management. The male-headed households and if HHH's primary occupation remains farming, it is expected to have positive impact on adoption of these practices because of having competitive aim of increasing production. Those who have rented in the land in addition to farming in their own land are expected to have negative impact on such adoption because they will be constrained with resources to make an investment in such technologies. Group membership is expected to increase the adoption rate

Table 6.1 Expected sign of socioeconomic variables against dependent variable of OCMPs

Independent variables	Definition and measurement	Dependent variables					References
		Mulch	Compost-shed	Bio-slurry	Bio-pesticides	Others	
Farm_method	Farmers practicing organic farming; 1 = yes, 0 otherwise	+ve	+ve	+ve	+ve	+ve	Own elaboration
HHHgender	Male-headed HH; 1 = yes, 0 otherwise	+ve	+ve	+ve	+ve	+ve	Own elaboration
HHHage	Age of HHH; in years	+ve/-ve	+ve/-ve	+ve/-ve	+ve/-ve	+ve/-ve	Akinola and Owombo (2012), Kuntariningsih and Mariyono (2013), and Grazhdani (2013)
HHHedu	Education of HHH; in years	+ve	+ve	+ve	+ve	+ve	Akinola and Owombo (2012), Grazhdani (2013), and Adolwa et al. (2010)
HHHprimary_occu	Primary occupation of HHH; 1 = farming, 0 otherwise	+ve	+ve	+ve	+ve	+ve	Own elaboration
Rent	Farmers renting by paying either cash or through crop sharing or mortgaging in farmland; 1 = yes, 0 otherwise	-ve	-ve	-ve	-ve	-ve	Own elaboration
Org_exp	Experience of practicing organic farming; in years	+ve/-ve	+ve/-ve	+ve/-ve	+ve/-ve	+ve/-ve	Kuntariningsih and Mariyono (2013) and Grazhdani (2013)
LFU	Labor force availability in HH; in labor force unit (LFU)	+ve/-ve	+ve/-ve	+ve/-ve	+ve/-ve	+ve/-ve	Akinola and Owombo (2012), Kuntariningsih and Mariyono (2013), and Mugwe et al. (2009)
LSU	Livestock holding in HH; in livestock unit (LSU)	-ve	+ve	+ve	+ve	+ve	Jaleta et al. (2012), Mugwe et al. (2009), and Adolwa et al. (2010)
Farm_size	Operational farm size; in ha	+ve	+ve	+ve	+ve	+ve	Akinola and Owombo (2012) and Grazhdani (2013)
Farm_income	Monetary and nonmonetary gross income from crops, livestock, and	+ve	+ve	+ve	+ve	+ve	Grazhdani (2013)

6.2 Socioeconomic Variables' Relation to Adoption of Organic Means of...

Variable	Description						Source
	selling trees; in Nepalese rupees (NRs.)/HH/year						Akinola and Owombo (2012) and Adolwa et al. (2010)
Non-farm_income	Income from non-farm activities (service, business, rent, remittance, pension and laboring); in NRs./HH/year	+ve/-ve	+ve/-ve	+ve/-ve	+ve/-ve	+ve/-ve	
Membership	Being in/formal group member formed for organic farming; 1 = yes, 0 otherwise	+ve	+ve	+ve	+ve	+ve	Own elaboration
Org_training	Organic farming-related training; number of times	+ve/-ve	+ve/-ve	+ve/-ve	+ve/-ve	+ve/-ve	Bizimana (2013)
VDC	Belonging to Phoolbari VDC; 1 = yes, 0 otherwise	+ve	+ve	+ve	+ve	+ve	Own elaboration
Agrovet	Distance to nearest store for agro-products; in km	+ve/-ve	+ve/-ve	+ve/-ve	+ve/-ve	+ve/-ve	Own elaboration
Market	Distance to nearest market; in km	+ve/-ve	+ve/-ve	+ve/-ve	+ve/-ve	+ve/-ve	Own elaboration
Credit	Credit taken for farming-related activities; 1 = yes, 0 otherwise	+ve	+ve	+ve	+ve	+ve	Own elaboration
Commercialization	Commercialization rate; in ratio of total quantity of crops sold/total produced	-ve	-ve	-ve	-ve	-ve	Own elaboration

because the whole purpose of such group is to enhance organic farming. Farmers belonging to Phoolbari VDC are also expected to have higher adoption rate as activities related to organic farming is more vibrant there compared to other two VDCs. Distance to agrovet and market could act as means of information on both organic and conventional technologies and provide accessibility to use them as well. Having access to credit is also expected to have positive impact as it enables farmers to make investments. Finally commercialization rate is anticipated to have negative impact on adoption as it mainly relies on conventional means for boosting production in short-run for higher profit.

6.3 Empirical Model

Various studies have adopted different models to assess factors influencing one's decision to adopt a certain practice. However, each model has its limitation that makes them inappropriate to be used for studies with certain purposes. For example, Heckman sample selection probit model has been used to analyze binomial choice of adopting conservation agriculture (Broeck et al. 2013) and perception and adaptation to climate change (Ndambiri et al. 2013) with respect to farmers' socioeconomic variables. However, this model does not differentiate between several kinds of conservation practices or perception and adaptation measures undertaken by households. Different farming practices can be impacted by different socioeconomic variables. Hence, to combine all sub-components into one and to assume impact of certain variable to have similar impact throughout may lead to biasness. When information of variables leading to a certain practice is known, it is clearer to seek out appropriate action to encourage adoption of one practice over another.

Univariate probit model provides another alternative of doing so by modeling each of the farming practices individually as a function of common set of explanatory variables. But it ignores unobserved and unmeasured common factors affecting different management practices. In other words, this model is undesirable for reason of failing to see relation among various management practices. Adopting any practices could be complementary or competing to each other. In this case, a farmer might use bio-pesticide and complement it with mulching. Likewise, bio-slurry might compete with compost as both require animal manure as a primary input. Thus, overlooking potential correlations among these practices may lead to statistical biasness and inefficient estimates (Nhemachena and Hassan 2007). Also since this study incorporates organic and conventional farmers, it would be interesting to see how these farmers differ in adopting such practices. Multinomial logit (MNL) model assumes independence of irrelevant alternatives (IIA) and that the practices be mutually exclusive which again in this case is not true. Farmer's decision to choose a certain practice might change when one or more additional alternative is available. Similarly, a farmer can choose two or more practices simultaneously. The drawback of multinomial discrete choice model is it fails to

6.3 Empirical Model

interpret effect of independent variables on adopting each practices separately (Golob and Regan 2002).

Considering the possibility of simultaneous adoption of soil management practices and potential correlations among these practices as well as between unobserved disturbances, multivariate probit (MVP) model has been used in this study. Furthermore, it relaxes the assumption of IIA. The MVP model assumes that given a set of explanatory variables, the multivariate response is an indicator of the event that some unobserved latent variable that is assumed to arise from a multivariate normal (Gaussian) distribution falls within a certain interval (Tabet 2007; Belderbosa et al. 2004). Referring to Tabet (2007), the MVP model in this study is characterized by a set of binary dependent variables Y_{ij}, where i is the independent observations and j available options of binary responses. Let Z_{ij} be a vector of latent variables so that:

$$Z_{ij} = \beta_0 + x_{ij}\beta + \varepsilon, \quad i = 1, \ldots, n \qquad (6.1)$$

where x_{ij} represents vector of explanatory variables which can be discrete or continuous, β_0 is coefficient of intercept, β is a vector of unknown parameters to be estimated, and ε is error term distributed as multivariate normal distribution with zero mean and unitary variance, $\varepsilon_i \sim N(0, \sum)$, where $\sum$ is variance-covariance matrix that has a value of one on the leading diagonal. Off-diagonal elements in the covariance matrix $\rho_{kj} = \rho_{jk}$ are an unobserved correlation between stochastic component of the kth and jth options (Young et al. 2009; Cappellari and Jenkins 2003). The relationship between Z_{ij} and Y_{ij} can be provided as follows:

$$Y_{ij} = \begin{cases} 1 & \text{if } Z_{ij} > 0; \\ 0 & \text{otherwise} \end{cases} \quad j = 1, \ldots, J \qquad (6.2)$$

By integrating over latent variables Z, the likelihood of observed discrete data can then be obtained by the following specification:

$$P(Y_{ij} = 1 | X_i, \beta, \Sigma) = \int A_{ij} \ldots \int A_{i1} \varnothing_T(Z_{ij} | X_i, \beta, \Sigma) \, dZ_{ij} \qquad (6.3)$$

where A_{ij} is interval $(0, \infty)$ if $Y_{ij} = 1$ and interval $(-\infty, 0]$ otherwise, and $\varnothing_T (Z_{ij} | X_i, \beta, \Sigma) \, dZ_{ij}$ is probability density function of standard normal distribution. To estimate MVP model, this study uses simulated maximum likelihood (SML) using Geweke-Hajivassiliou-Keane (GHK) simulator, which is considered as the most popular method for evaluating multivariate normal distribution function in STATA developed by Cappellari and Jenkins (2003). According to Cappellari and Jenkins (2003), when number of draws and observations are infinite, SML estimator is consistent. Simulation (finite sample) bias can be reduced to negligible levels when number of draws is raised with the sample size. Generally for large sample size (thousand and above), it is sufficient to have number of draws equal to square root of the sample size. But for small sample size, number of draws should be sufficiently large. Thus, for this study, number of draws (R) was set to 100 (from default of $R = 5$) to ensure reliable estimates.

As per the regression rule, diagnostic tests were carried out. For each individual choice variables, OLS estimates were run against the same set of explanatory variables to conduct diagnostic tests in order to check if there is any problem of multicollinearity and heteroscedasticity in the data. The variation inflation factor (VIF) value for all the independent variables was much below 10, which means that there is no problem of multicollinearity among the variables. Likewise, Breusch-Pagan/Cook-Weisberg showed significant P-value for all individual choice variables, at varying level of significance, thus rejecting null hypothesis of homoscedasticity. It means that there are linear forms of heteroscedasticity. On the other hand, White's test did not show significant P-value for any of the individual choice variables, implying that there is no problem of nonlinear forms of heteroscedasticity, i.e., the variance of error term is constant. To overcome the problem heteroscedasticity of any kind, model estimation was conducted using robust standard errors.

The empirical model for the best fit model, generated after removing redundant variables (with insignificant P-value) through backward elimination method, can be given by:

$$\begin{aligned} y_{n=5} = \ & \beta_0 + \beta_1 \ \text{farm\_method} + \beta_2 \ \text{rent} + \beta_3 \ \text{org\_exp} \\ & + \beta_4 \ \text{LFU} + \beta_5 \ \text{LSU} + \beta_6 \ \text{farm\_size} + \beta_7 \ \ln\_\text{farm\_income} \\ & + \beta_8 \ \ln\_\text{non} - \text{farm\_income} + \beta_9 \ \text{org\_training} \\ & + \beta_{10} \ \text{credit} + \beta_{11} \ \text{commercialization} + \varepsilon \end{aligned} \quad (6.4)$$

where y = mulch, compost-shed, bio-slurry, bio-pesticide, and/or others (plastic cover and/or vermicompost) and ln is natural log.

6.4 Organic Means of Crop Management Practices

Figures 6.6 and 6.7 provide information on the extent of adoption of OCMPs and its distribution across the two farming systems prevalent in the study areas. About 72% of the respondents did mulching, 23% have a compost-shed, 43% have biogas from which they get bio-slurry, 19% uses bio-pesticides, and only 4% and 6% use plastic cover and vermicompost, respectively. Among others, mulching is the most traditional form of soil fertility management practice considered for this study, which is why its adoption rate is higher than other practices. Mostly organic farmers compared to conventional farmers adopt all OCMPs considered in this study. Especially in case of practicing compost-shed and bio-pesticides, the share of organic farmers is higher by double compared to conventional farmers. In case of conventional farmers, however, they relied on chemical fertilizers such as urea, DAP, and MOP; chemical pesticides such as insecticide, weedicide, and herbicide; and micronutrients such as zinc, boron, vitamin, and hormones.

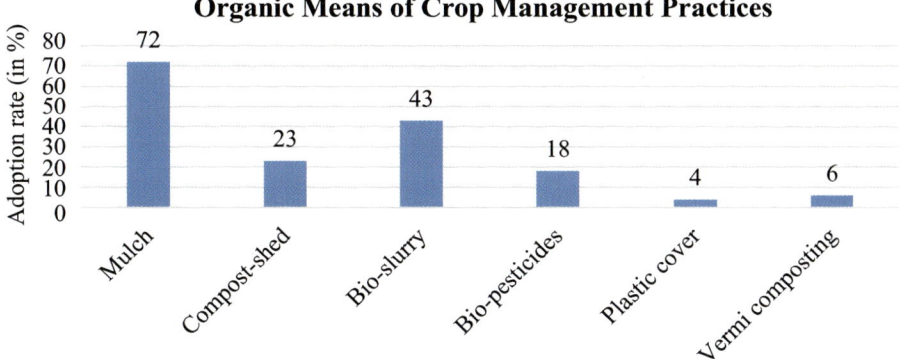

Fig. 6.6 Organic means of crop management practices adopted by the respondents. *Source*: Field survey (2013)

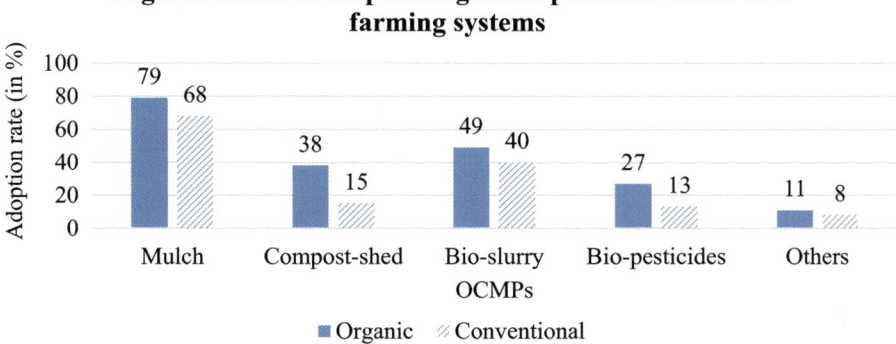

Fig. 6.7 Organic means of crop management practices adopted across two farming systems. *Source*: Field survey (2013)

6.5 Socioeconomic Impact on Adoption of Organic Means of Crop Management Practices

The result from MVP model is presented in Table 6.2. Likelihood ratio statistics as denoted by Wald $\chi2$ is highly significant ($p = 0.0000$), which shows goodness of fit, i.e., variables sufficiently explain the model. Also likelihood ratio test of null hypothesis of the absence of correlation among individual equations is strongly rejected ($p = 0.0000$), thus justifying rationale to estimate all equations simultaneously using MVP rather than estimating individually. Some of the directions of signs of independent variables are as per the expectation while others are not. Most importantly, a single variable does not have similar direction of impact across all dependent variables, as was presumed to be.

Table 6.2 Parameter estimates of multivariate probit model for organic means of crop management practices

Explanatory variables	Mulch	P-value	Compost-shed	P-value	Bio-slurry	P-value	Bio-pesticide	P-value	Others	P-value
farm_method	0.41	0.078*	0.58	0.016**	0.22	0.375	−0.01	0.959	0.07	0.829
Rent	−0.25	0.216	0.18	0.402	−0.48	0.015**	0.42	0.061*	−0.02	0.940
Org_exp	−0.01	0.524	0.01	0.656	−0.02	0.282	−0.02	0.305	0.00	0.797
LFU	−0.02	0.583	0.07	0.164	−0.06	0.283	0.04	0.405	0.01	0.834
LSU	0.02	0.698	0.01	0.882	0.05	0.305	0.06	0.291	−0.03	0.530
Farm_size	0.22	0.348	0.40	0.100*	0.06	0.803	−0.10	0.681	0.06	0.853
ln_farm_income	0.21	0.055*	0.18	0.127	0.67	0.000***	−0.09	0.506	0.11	0.462
ln_non-farm_income	0.01	0.507	0.01	0.492	−0.01	0.769	0.02	0.482	0.01	0.508
Org_training	0.03	0.621	0.06	0.201	0.14	0.014**	0.32	0.000***	0.03	0.671
Credit	0.95	0.009***	0.38	0.194	−0.35	0.218	0.21	0.450	0.46	0.105
Commercialization	−0.03	0.835	0.04	0.781	0.02	0.897	−0.08	0.607	0.18	0.261
Constant	−2.20	0.080*	3.93	0.005***	−8.03	0.000***	−0.72	0.634	−3.07	0.086*
Correlation coefficients			Coefficient			P-value				
$\hat{\rho}_{21}$			0.31			0.006***				
$\hat{\rho}_{31}$			−0.19			0.069*				
$\hat{\rho}_{41}$			−0.06			0.656				
$\hat{\rho}_{51}$			0.25			0.053*				
$\hat{\rho}_{32}$			−0.01			0.914				
$\hat{\rho}_{42}$			0.02			0.901				
$\hat{\rho}_{52}$			0.13			0.377				
$\hat{\rho}_{43}$			0.07			0.591				
$\hat{\rho}_{53}$			0.61			0.000***				
$\hat{\rho}_{54}$			0.47			0.000***				

6.5 Socioeconomic Impact on Adoption of Organic Means of Crop Management Practices

Draws	100
Number of observations	285
Wald χ^2 (55)	222.73
P-value	0.0000***
Log pseudo likelihood	−623.87156
Likelihood ratio test rho21 = rho31 = rho41 = rho51 = rho32 = rho42 = rho52 = rho43 = rho53 = rho54 = 0, χ^2 (10) = 42.2805, P-value = 0.0000***	

Source: Field survey (2013)

Note: ***1%, **5%, and *at 10% level of significance

Compared to conventional farmers, organic farmers have higher probability of mulching, significant at 10%, and constructing compost-shed, significant at 5% level. Mulching is the most traditional way of farming which is why more of organic farmers are inclined to using it. Most of the farmers received partial funding (25% of the total cost) from a local NGO to construct compost-shed. Fund eligibility depends on financial ability of a farmer to supply the remaining cost and active participation in group activities. Organic farmers in the study areas have received more training as shown by previous study (Chap. 3), indicating active participation in the group activities, and so they are prioritized to be recipient of such fund. However, only two farmers are selected per year, and some farmers are not in a position to even finance the rest of the fund, which has slowed down the adoption rate.

Those who have rented farmland in addition to farming in their own land, their possibility of using bio-slurry decreases, significant at 5%, while using bio-pesticides increases, significantly at 10%. Tenant farmers are usually resource poor which is why their chances of constructing biogas that requires higher initial investment are less. Since bio-pesticides could be made from resources that are available in the farm, it could the reason be why tenant farmers' probability of using it increases, as it is cheaper than buying commercially available pesticides in the market. Higher farm size results in higher propensity of constructing compost-shed. Constructing compost-shed too requires higher initial investment, and as explained earlier, poor farmers are not able to construct one even if they are offered partial cost. Farm size indicates higher resource holding which could have increased their chances of constructing compost-shed. Those who have higher farm income, their probability of mulching and using bio-slurry increases, significant at 10% and 1%, respectively. Farm income indicates producing more crops, the residues of which could be used for mulching. Like farm size, farm income also signifies higher resource endowment or capital availability, thus increasing ability of installing biogas or even mulching if it is done using plastic, which requires more investment.

Training has positive impact on using bio-slurry and bio-pesticide, significant at 5% and 1%, respectively. Importance of bio-slurry and bio-pesticide is very much promoted through training conducted from the formed groups. Experience of organic farming system has negative impact on using bio-pesticide. Farmers often perceive bio-pesticides to consume time and requiring much labor during preparation and application as well. For example, it takes about 2 weeks or more to prepare bio-pesticides depending on the amount of sunlight it receives. During this time, the problem of pests and disease already would have multiplied, and its impact is also perceived to be slower compared to conventional means. Further they have to replace plastic container in which to prepare bio-pesticides every few years, as its life span is very short. These reasons might have contributed to its lower adoption.

Credit increases probability of adopting mulching, significant at 1%. Though credit might not have been used directly for mulching, it can have counter-effect through other activities which farmers actually used it for. For example, using credit for cultivating more crops results in more crop residue for mulching. Estimated correlation coefficient among various OCMPs is significant for five out of ten combinations. Mulching and compost-shed are positively related and are highly significant at 1%. This means that farmers are combining these practices for soil management though both compete for crop residue. Mulching and bio-slurry are negatively related, significant at 10%. It indicates that mulching and livestock, which ultimately provide manure for biogas, compete indirectly for crop residue, thus having negative impact. Farmers are incorporating uncommon ways of soil management practices such as plastic cover and/or vermicompost with bio-slurry and bio-pesticide, significant at 1%, or even with traditional practice such as mulching, significant at 10% level. This means that those farmers who are already practicing some ways of soil or pest management practices are also incorporating not so common practices such as plastic cover and/or vermicompost.

6.6 Summary

Even though there is influx of modern inputs like chemical fertilizers, pesticides, and micronutrients, conventional farmers still incorporate organic means of soil fertility and pest management practices such as mulching, compost-shed, bio-slurry, bio-pesticides, and plastic cover and/or vermicompost. However, adoption rate for all of such practices is high among organic farmers, indicating that organic farmers are keener on practicing various ways of sustainable soil fertility and pest management practices, especially when it comes to mulching and constructing compost-shed. Mulching is the most traditional way of soil management practice and has higher prospect to be adopted by organic farmers as they mainly follow traditional way of farming, those having higher farm income as they will also have produced more crops which further provides more crop residue for mulching, and those who have taken credit which might not have direct impact but credit for higher investment in crop cultivation results in higher crop residue available for mulching. However in some instances, adoption can be hindered by lack of fund, such as in the case of compost-shed. Thus, it is advisable that fund assistance should be increased so as to increase adoption rate of compost-shed by the majority. Tenant farmers have less resource holding which is why their probability of constructing biogas decreases, as it requires higher initial investment. Similarly those who have bigger farm size or higher farm income indicates being resource rich, and thus their chances of adopting higher investment requiring practices such as compost-shed and biogas, respectively, too increase. This further proves that financial ability is the major drawback for adoption of these sustainable practices. One of the ways to increase the adoption rate is training as it complements technical knowledge required to implement these practices. Also if such practices largely rely on locally available resources such as bio-pesticides, then even farmers facing financial constraint can adopt these practices such as tenant farmers.

Farmers also tend to complement most of the practices. They are practicing uncommon techniques such as plastic cover and/or vermicompost alongside biogas, bio-pesticides, or even with traditional ones like mulching. Thus, any additional organic means of soil fertility or pest management practices can be introduced to those households who are already adopting one of such practices as they tend to be more aware and enthusiastic on adopting such practices. But sometimes, these practices become substitutes because of their nature of relying on the same input such as mulching and biogas directly or indirectly depending on crop residue. Thus, any effort to enhance such adoption rate can consider these characteristics of various practices. Hence, adoption of organic means of crop management practices is influenced in different ways by various socioeconomic factors, which should be regarded before any intervention.

References

Adolwa IS, Esilaba AO, Okoth P, Mulwa MR (2010) Factors influencing uptake of integrated soil fertility management knowledge among smallholder farmers in Western Kenya. 12th KARI biennial scientific conference: transforming agriculture for improved livelihoods through agricultural product value chains. Africa Soil Information Service (AfSIS), Nairobi, pp 1146–1152

Akinola A, Owombo P (2012) Economic analysis of adoption of mulching technology in yam production in Osun State, Nigeria. Int J Agricult Forestry 2(1):1–6

Belderbosa R, Carree M, Diederen B, Lokshin B, Veugelers R (2004) Heterogeneity in R&D cooperation strategies. Int J Ind Organ 22:1237–1263

Bhat BR, Ghimire R (2008) Promotion of organic vegetable production through farmers' field school in Chitwan, Nepal. 16th IFOAM Organic World Congress, International Federation of Organic Agriculture Movement (IFOAM), Modena

Bista P, Ghimire R, Shah CS, Pande KR (2010) Assessment of soil fertility management practices and their constraints in different geographic locations of Nepal. Forum Geografic 9:41–48

Bizimana C (2013) Strategies for improving adoption of soil-fertility technologies in Rwanda. Reg Strat Anal Knowl Support Syst (ReSAKSS) 19:1–7

Broeck GV, Grovas RR, Maertens M, Deckers J, Verhulst N, Govaerts B (2013) Adoption of conservation agriculture in the Mexican Bajío. Outl Agricult 42(3):171–178

Cappellari L, Jenkins SP (2003) Multivariate probit regression using simulated maximum likelihood. Stata J 3(3):278–294

EPA (2013) Biopesticides. United States Environmental Protection Agency (US-EPA), Washington, DC

Golob TF, Regan AC (2002) Trucking industry adoption of information technology: a structural multivariate discrete choice model. Transport Res Part C Emerg Technol 10:205–228

Grazhdani D (2013) An analysis of factors affecting the adoption of resource conserving agricultural technologies in Al-Prespa park. Natura Montenegrina 12(2):431–443

Green VS, Brye KR (2008) Soil quality: an essential component of environmental sustainability. Electron J Integrat Biosci 6(1):1–2

Huili G, Dan M, Xiaojuan L, Feng Z (2013) Soil degradation and food security coupled with global climate change in Northeastern China. Chin Geogr Sci 23(5):562–573

IFDC (2014) Integrated soil fertility management (ISFM). International Fertilizer Development Center (IFDC). http://www.ifdc.org/Technologies/ISFM/. Retrieved 2 May 2014

Jaleta M, Kassie M, Shiferaw B (2012) Tradeoffs in crop residue utilization in mixed crop-livestock systems and implications for conservation agriculture and sustainable land management. International Association of Agricultural Economists (IAAE) Triennial Conference, Foz do Iguacu, pp 18–24

References

Kabuli A, Phiri M (2007) Farmer perceptions, choice and adoption of soil management technologies in maize-based farming systems of Malawi. Food and Agriculture Organization of the United Nations (FAO), Rome

Kuntariningsih A, Mariyono J (2013) Socio-economic factors affecting adoption of hybrid seeds and silver plastic mulch for chili farming in Central Java. SEPA 9(2):297–308

Montri A, Biernbaum J (2009) Management of the soil environment in high tunnels. HortTechnology 19(1):34–36

Mugwe J, Mugendi D, Mucheru-Muna M, Merckx R, Chianu J, Vanlauwe B (2009) Determinants of the decision to adopt integrated soil fertility management practices by smallholder farmers in the Central Highlands of Kenya. Expl Agric 45:61–75

Ndambiri HK, Ritho CN, Mbogoh SG (2013) An evaluation of farmers' perceptions of and adaptation to the effects of climate change in Kenya. Int J Food Agricult Econom 1(1):75–96

Nhemachena C, Hassan R (2007) Micro-level analysis of farmers' adaptation to climate change in Southern Africa. Int Food Policy Res Inst (IFPRI), Washington, DC

OECD (2008) Natural resources and pro-poor growth: the economics and politics. Organisation for Economic Co-operation and Development (OECD), Paris

OEFFA (n.d.) OEFFA organic certification fact sheet: compost and manure. Ohio Ecological Food and Farm Association (OEFFA), Ohio

Schonbeck M (2012) Organic mulching materials for weed management. eOrganic, Missouri

Shrestha N, Raes D, Sah SK (2013) Strategies to improve cereal production in the Terai region (Nepal) during dry season: simulations with aquacrop. Procedia Environ Sci 19:767–775

Tabet A (2007) Bayesian inference in the multivariate probit model: estimation of the correlation matrix. University of British Columbia, Vancouver

Watson C, Atkinson D, Gosling P, Jackson L, Rayns F (2002) Managing soil fertility in organic farming systems. Soil Use Manag 18:239–247

Young G, Valdez EA, Kohn R (2009) Multivariate probit models for conditional claim-types. Insurance Math Econom 44(2):214–228

Chapter 7
Crop Diversification Under Organic and Conventional Farming Systems

Abstract This chapter analyzes crop diversification between organic and conventional farming systems using Shannon Diversity Index (SHDI). It captures both richness (number) and evenness (abundance) of crops and analyzes impact of livelihood assets on it using ordinary least square model. Organic farming system in the study areas is richer in integrating more number of crop types (richness) but is poor in evenness, which resulted in having lower SHDI than conventional farming system. Since crop evenness is better indicator of improved productivity than crop richness, it can be implied that farmers, especially organic farmers, should be made aware of this fact in order to improve their overall productivity. The socioeconomic variables that have significant positive impact on SHDI are education attainment, livestock holding, non-farm income, group membership, training, and farther distance to the market.

7.1 Introduction

Crop diversification means diversifying number of crops grown in a particular area at any given time. In other words, it is contrary to mono-cropping that focuses on growing a single crop. Crop diversification could be either horizontal diversification, which is adding or substituting crops into the current cropping method, or vertical diversification that includes value addition activities that generally occur in industrialization stage. This chapter considers horizontal diversification of crops that is commonly practiced by many countries in Asia-Pacific region (FAO 2001). Crop diversification can provide numerous benefits on environmental, social, and economic grounds. It is an indication of biodiversity as with diverse crops, the number of habitats is also expected to increase. It leads to intercropping that delivers ecosystem services of improved pest control, resource use efficiency through facilitation and complementarity between species, nutrient cycling processes and product quality, and lower level of weed infestation and nitrate leaching compared to single cropping.

One of the most common ways is intercropping cereals with legumes as latter is known to supply nutrients (maize + cowpea, maize + soybean, maize + pigeon pea, maize + ground nuts, maize + soybean, sorghum + cowpea, millet + groundnuts, and rice + pulses). In study areas too, farmers are found intercropping for various reasons. For example, they intercropped crops needing more sunlight with less needing ones (wheat + rapeseed), crops needing more inputs with less needing ones (wheat + peas; corn + cowpea; chilli or soybean; cauliflower + coriander, radish, onion, carrot, or garlic; potato + field beans), deep rooted with shallow-rooted crops (pigeon pea + green gram), long duration with short duration crops (corn + soybean, carrot + radish), and for pest management (cauliflower + coriander, cabbage + garlic, spinach + onion, garlic, or coriander).

Crop diversification will help maintain production over time without degrading the environment. It improves resilience to withstand stress and other disturbances such as during incidences of single crop failure, environmental adversity, or socioeconomic shocks. Diversified farm can supply various combinations of nutrients consequently enriching households' diet and reducing household market dependency. In fact, it can develop market orientation by producing high-value crops that generate additional income. It also improves employment opportunities by cultivating crops that have varying growing season all year round (Andersen 2005; Scialabba 2007; Johnston et al. 1995; UNCTAD 2003; Sipiläinen et al. 2008; Matusso et al. 2012).

While crop diversification provides numerous benefits, the diversity of enterprises may result in lower efficiency (Belicka 2005; FAO 2014). Because of this, farmers try to reduce the number of crops as much as possible for economic incentive. Organic farmers are known to rely on the natural services provided by such crop diversification, while conventional farmers rely on chemical inputs for production maximization. While it is a common pattern that conventional farmers have less number of crops integrated into their farming system, especially in developed countries, it is not so clear in case of developing countries where they still follow traditional ways complemented with modern inputs (Bista et al. 2010; Kabuli and Phiri 2007). This chapter assesses to what extent crop diversification differs among farmers practicing organic and conventional farming systems. Measuring crop diversity is not only limited to a farming system, but studies have expanded its horizon to socioeconomic impact as well (Gauchan et al. 2005; Pandey 2013; Benin et al. 2003; Rehima et al. 2013). Thus, it also aims to understand how socioeconomic background, unique to each farmer, impacts crop diversification.

7.2 Socioeconomic Variables' Relation to Crop Diversification

Based on various literature reviews, Table 7.1 provides the expected direction of each of the variables used with respect to the crop diversity. Organic farming is known to have more diverse crops than conventional farming as diversity harmonizes with the basic principles of crop rotation or intercropping under organic farming that helps to function and manage farm through natural means (UNEP-UNCTAD 2008; UNCTAD 2003; Bengtsson et al. 2005). Since the basic intention

7.2 Socioeconomic Variables' Relation to Crop Diversification

Table 7.1 Expected relation of explanatory variables with respect to dependent variable of SHDI

Variables	Definition and measurement	Expected sign	References
Farm_method	Farmers practicing organic farming; 1 = yes, 0 otherwise	+ve	UNEP-UNCTAD (2008), UNCTAD (2003), and Bengtsson et al. (2005)
HHHgender	Male-headed HH; 1 = yes, 0 otherwise	-ve	Own elaboration
HHHage	Age of HHH; in years	+ve	Gauchan et al. (2005)
HHHedu	Education of HHH; in years	+ve	Gauchan et al. (2005), Pandey (2013), Benin et al. (2003), and Rehima et al. (2013)
HHHprimary_occu	Primary occupation of HHH; 1 = farming, 0 otherwise	+ve/-ve	Own elaboration
Rent	Farmers renting by paying either cash or through crop sharing or mortgaging in farmland; 1 = yes, 0 otherwise	+ve/-ve	Own elaboration
Org_exp	Experience of practicing organic farming; in years	+ve	Own elaboration
LFU	Labor force availability in HH; in labor force unit (LFU)	+ve	Gauchan et al. (2005)
LSU	Livestock holding in HH; in livestock unit (LSU)	+ve/-ve	Gauchan et al. (2005), Benin et al. (2003), and Rehima et al. (2013)
Farm_size	Operational farm size; in ha	+ve/-ve	Gauchan et al. (2005), Pandey (2013), and Rehima et al. (2013)
Non-farm_income	Income from non-farm activities (service, business, rent, remittance, pension and laboring); in NRs./HH/year	+ve	Gauchan et al. (2005)
Membership	Being informal/formal group member formed for organic farming; 1 = yes, 0 otherwise	+ve	Rehima et al. (2013)
Org_training	Organic farming-related training; number of times	+ve	Rehima et al. (2013)
VDC	Belonging to Phoolbari VDC; 1 = yes, 0 otherwise	+ve	Own elaboration
Agrovet	Store selling agro-products (seeds, fertilizers, pesticides, equipment, veterinary medicine, etc.); distance to nearest agrovet (in km)	+ve/-ve	Own elaboration
Market	Market to selling agro-products; distance to nearest market (in km)	+ve	Gauchan et al. (2005)
Credit	Credit taken for farming-related activities; 1 = yes, 0 otherwise	-ve	Own elaboration

of conventional farming is to reduce cost and increase production to maximize profit that could be achieved through crop specialization, it could be implied that crop diversity in this farming system tends to decrease with increasing intensity of specialization.

Households' characteristics like age of decision-makers resulted in increased diversity which could be related to their experience or their unwillingness to consider or accept modern variety that demands specialization. Education of household head and household labor availability is also expected to expand the variety choice (Gauchan et al. 2005). Educated farmers become aware of various crops through better interaction, and their ability to understand and utilize technical information associated with new crops also improves. They are also knowledgeable about nutritional value of various crops, which leads to crop diversification (Pandey 2013; Benin et al. 2003; Rehima et al. 2013). Labor availability will lead to increased diversity if there are less non-farm opportunities. Wealth-related variables like farm size, livestock holding, non-farm income, and the like could also improve diversity richness because of their ability to take more risk (Gauchan et al. 2005). Studies also show that resource-rich farmers with larger farm size are cultivating less varieties of crops through specialization as per the market demand or because it becomes difficult to manage multiple crops on larger area (Pandey 2013; Rehima et al. 2013).

As for those with small farm size, crop diversity intensifies to meet household food requirement, but a very small farm might not necessarily result in same way, as farmers will find it feasible to focus on other income sources. Studies by Benin et al. (2003) and Rehima et al. (2013) showed that livestock holding might actually work as a safety net against crop production failure, and thus farmers with higher livestock might lead to greater specialization, which means lesser diversification. Distance from market which is also expected to have positive and higher sales of particular variety has negative relation to diversity (Gauchan et al. 2005). It is because with easier access to market, households will definitely take its advantage by specializing crops with higher market value. Those with poor market access are more likely to rely on diversification to meet their consumption needs and to avoid transaction costs of going to the market to buy their food. In the same way, extension services positively contribute to crop diversity, as it is associated with spread and adoption of new technologies that could be relevant to diversification (Rehima et al. 2013). Group membership and training are expected to contribute positively in this regard as it concerns with organic farming.

It is expected that females will understand more about the benefit of having diversified crop as a requirement for household consumption as they are the ones who take charge of feeding their family. Farming as a primary occupation of HHH may or may not lead to higher diversification because on the one hand such farmer will think about fulfilling his family's consumption requirement first but on the other hand, he may be too competitive in earning higher income resulting in cultivating few crops with commercial value. Those farmers who have rented in the farmland in addition to farming in their own land could also have similar impact on crop diversification. Experience of practicing organic farming and farming in

Phoolbari VDC, which is more vibrant in performing activities related to organic farming, is also projected to have positive relation. This study also takes distance to nearest agrovet, which might act as a medium of information and service provider, which again could be inclined toward or deviate from organic farming. Finally, because of having used credit for investing in commercial crops in addition to other purposes, it is expected to have negative relation to diversity.

7.3 Empirical Model

The study calculates crop diversity using Shannon Diversity Index (SHDI). It is also commonly known as Shannon-Wiener/Weiner/Weaver Index. It captures both richness and evenness of species diversity. Richness implies the number of species cultivated, whereas evenness refers to how evenly the cultivated area is distributed to various species. Species richness is the simplest way to measure diversity, but evenness captures a broader picture by taking relative abundance of species that enriches diversity. It is not necessary that species richness and evenness always have a positive relation. In many cases, diversity can be altered by changes in evenness without any change in species richness. Therefore, species richness and evenness should be assumed as two independent indices (Zhang et al. 2012). Wilsey and Potvin (2000) found that species evenness has more linear relationship with total productivity than with species richness. Including these two indices can give better understanding of the status of diversity. SHDI has been used in different studies for assessing diversity of numerous kinds (Sipiläinen et al. 2008; Edesi et al. 2012).

SHDI can be calculated with the following formula:

$$H = \sum_{i=1}^{n} (P_i * \ln P_i) \tag{7.1}$$

where H is SHDI, n is number of cultivated crops, P is share of area occupied by crop i from total cultivated area, and ln is natural logarithm. The value usually comes in negative in which case it should be converted to positive sign. When diversity index equals zero, it suggests that there is only one crop and hence no diversity. Thus, the value increases with number of crops cultivated and when the area under which various crops cultivated becomes more evenly distributed. This study uses six different categories of crops with numerous types under each to calculate this index (Appendix II).

In order to assess impact of socioeconomic variables of farmers on crop diversity, ordinary least square (OLS) model has been used. It is the most frequently used model for fitting the regression line (Hoffmann 2010). OLS model can be expressed as:

$$y_i = \beta_0 + x_i\beta_i + \varepsilon_i \quad (7.1)$$

where y = SHDI, x = HH's socioeconomic characteristics, i = number of observations, β_0= coefficient of intercept, β_i = parameter to be estimated, and ε = error term.

The empirical specification for the model, generated after removing redundant variables (with insignificant P-value) through backward elimination method, can be given by:

$$\begin{aligned} \text{SHDI} = {}& \beta_0 + \beta_1 \text{ farm\_method} + \beta_2 \text{ HHHedu} + \beta_3 \text{ org\_exp} \\ & + \beta_4 \text{LFU} + \beta_5 \text{ LSU} + \beta_6 \text{ farm\_size} + \beta_7 \text{ ln\_non-farm\_income} \\ & + \beta_8 \text{ membership} + \beta_9 \text{ org\_training} + \beta_{10} \text{ VDC} \\ & + \beta_{11} \text{ agrovet} + \beta_{12} \text{ market} + \beta_{13} \text{ credit} + \varepsilon \end{aligned} \quad (7.3)$$

where ln is natural log.

Variation inflation factor (VIF) at 1.42 is well below 10, indicating multicollinearity among the variables does not exist. Likewise, Breusch-Pagan/Cook-Weisberg gave significant P-value (0.0359), thus rejecting null hypothesis of homoscedasticity and implying that there are linear forms of heteroscedasticity. White's test did not show significant P-value (0.3818), suggesting that there is no problem of nonlinear forms of heteroscedasticity, i.e., variance of the error term is constant. To correct heteroscedasticity of any kind, model estimation was conducted using robust standard errors.

7.4 Crop Diversity in Organic and Conventional Farming Systems and Influence of Socioeconomic Factors

Figure 7.1 shows distribution of number of crop types under each category across the two farming systems. It is found that the number of crop types under all crop categories (cereals, vegetables, spices, pulses, oil seeds, and fruits) is lesser in conventional farming compared to organic farming. Therefore, it can be implied that the overall crop types or richness is higher in organic farming than conventional farming. Table 7.2 provides result from OLS model for SHDI against various HH socioeconomic variables. The P-value for the regression model as a whole is highly significant at 1%, which supports the existence of relationship of independent variables with dependent variable. The R^2 value suggests that about 27% of the total variation in value of dependent variable is explained by independent variables in this regression equation.

The result shows that organic farming has lower SHDI than conventional farming by 0.17, which is highly significant at 1%. It could be understood that unlike the traditional belief, organic farming might not necessarily be better in SHDI than conventional farming. Moreover, organic farming has higher number of

7.4 Crop Diversity in Organic and Conventional Farming Systems and Influence...

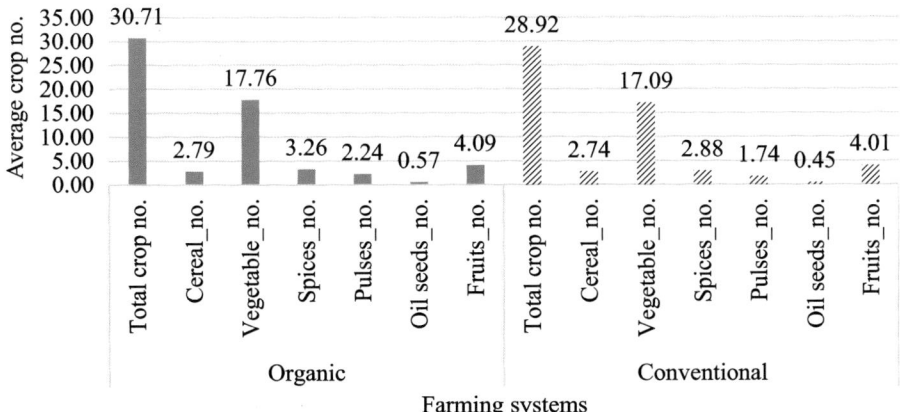

Fig. 7.1 Distribution of crop types under various categories and across two farming systems (*Source*: Field survey 2013)

Table 7.2 Result from ordinary least square model for Shannon Diversity Index

Variables	Coefficient	P-value
Farm_method	−0.17	0.004***
HHHedu	0.01	0.001***
Org_exp	0.004	0.149
LFU	0.01	0.289
LSU	0.03	0.007***
Farm_size	−0.01	0.805
ln_non-farm_income	0.01	0.024**
Membership	0.10	0.081*
Org_training	0.04	0.002***
VDC	0.17	0.000***
Agrovet	0.01	0.275
Market	0.01	0.032**
Credit	−0.04	0.523
Constant	2.65	0.000***

Source: Field survey (2013)
Note: ***1%, **5%, and *at 10% level of significance
Number of observations = 285, Prob >F = 0.0000***
$F(13, 271) = 8.29$, $R^2 = 0.2710$, Root MSE = 0.33573

crops (Fig. 7.1) but lower SHDI, which indicates that the crops cultivated are not as evenly distributed as in conventional farming.

Education of HHH tends to contribute more in crop diversity. A year increase in formal education will have index increase by 0.01, significant at 1%. As referred in previous studies, educated farmers are better aware of various crops through interaction, have better comprehension of various crop types, and probably are more knowledgeable about nutritional value of such crops. A year increase in experience in organic farming increases SHDI by 0.004, suggesting that with such experience, farmers' realization of benefits of crop diversity also increases. A unit increase in LFU increases SHDI by 0.01, signifying higher diversity demands higher labor for managing diverse crops. A unit increase in LSU increases SHDI by 0.03, which is significant at 1%. Livestock holding complements crop diversity as a way of traditional integrated farming which could be the reason why these variables complement each other.

A hectare increase in farm size is negatively correlated with SHDI by 0.01, entailing that with bigger farm size, it will be difficult to have diversified farm, which demands more labor and other resources. Thus, at some point, farmers intensify certain crops rather than keep adding other crop types due to difficulty in managing larger farms. Non-farm income is positively related to increasing SHDI at 5% significance. It suggests that as income from non-farm source increases by a percent, it will increase SHDI by 0.01. With non-farm income source, farmers would no longer have to be competent in crop specialization; the intention of which is to increase income. Thus, they would rather grow lesser amount but with more varieties to fulfill their own household need.

Membership in a group established for the purpose of organic farming and training provided through it has positive correlation in increasing SHDI by 0.10 and 0.04, significant at 10% and 1%, respectively. This implies that crop diversity is encouraged through the activities including training and interaction that takes place in these groups. It has been able to instill basic features of organic farming through higher SHDI. Farming in Phoolbari VDC is found to have higher SHDI by 0.17 on average than in other two VDCs, significant at 1%. A kilometer distance to agrovet will increase SHDI by 0.01, indicating that inconvenient access to various agro-inputs which are offered by agrovets, especially chemical inputs, encourages farmers to incorporate more number of crops with better distribution of each crop rather than specializing in few. This also indicates being vigorous in multi-cropping as a way to manage pests, an alternative to using chemical pesticides. A kilometer distance to market will increase SHDI by 0.01, significant at 5%. As observed by Gauchan et al. (2005), easier access to market will encourage households to specialize in those crops which have higher market value. Contrarily, poor market access makes farmers more likely to diversify the crop to meet their own consumption need and to avoid transaction cost required for buying or selling the crops. Finally credit has negative impact on SHDI by 0.04, which means that among others, farmers' use of credit for commercial farming leads to decrease in SHDI.

7.5 Summary

Diversity richness and evenness are two separate entities, the latter being a more comprehensive way of measurement. Organic farming in the study area is richer in integrating more number of crop types (richness) but is poor in evenness, which resulted in having lower SHDI than conventional farming. Thus, organic farming should be focused more on this aspect, which literatures indicate will lead to more balanced and enriched biodiversity, thus improving the environmental services.

Socioeconomic variables that have significant positive impact on SHDI are education attainment, livestock holding, non-farm income, group membership, training, and belonging to Phoolbari VDC. Clearly, educated farmers have more knowledge on various crops and its benefits to health. Non-farm income allows farmers to intensify diversification for own household consumption rather than having to specialize for increasing income. Membership in a group formed for the purpose of organic farming and training related to organic farming can improve SHDI because the purpose of such group formation and training is to make farmers aware of benefits of agro-ecological principles resulting in improvement of soil fertility and hence the production. Farming in Phoolbari VDC has better SHDI, which means that other VDCs should be focused more in improving the SHDI. Finally farther the distance to the market will encourage farmers to increase SHDI because they will prioritize on being self-sufficient and avoid buying or selling in the market to save the transportation cost. Easier access to market leading to low SHDI suggests that market is only favorable for few selected crops, which will encourage farmers for crop specialization. If there were such opportunities for variety of crops, it might lead to diversifying crops, which is also beneficial for overall production through various environmental services. Therefore, any effort to improve SHDI should consider these characteristics. Most importantly the effort should also be made to cultivate crops more evenly in addition to having numerous types to reap more benefit from environmental point of view, ultimately resulting in higher production.

References

Andersen MK (2005) Competition and complementarity in annual intercrops – the role of plant available nutrients. The Royal Veterinary and Agricultural University. Department of Soil Science, Copenhagen

Belicka I (2005) Organic food: ongoing general aspects. Environmental friendly food production system: requirements for plant breeding and seed production (ENVIRFOOD), Latvia

Bengtsson J, Ahnstrom J, Weibull A-C (2005) The effects of organic agriculture on biodiversity and abundance: a meta-analysis. J Appl Ecol 42(2):261–269

Benin S, Smale M, Gebremedhin B, Pender J, Ehui S (2003) The determinants of cereal crop diversity on farms in the Ethiopian highlands. 25th international conference of agricultural economists, Durban, pp 1–26

Bista P, Ghimire R, Shah CS, Pande KR (2010) Assessment of soil fertility management practices and their constraints in different geographic locations of Nepal. Forum Geografic 9:41–48

Edesi L, Jarvan M, Adamson A, Lauringson E, Kuht J (2012) Weed species diversity and community composition in conventional and organic farming: a five-year experiment. Žemdirbystė Agricult 99(4):339–346

FAO (2001) Crop diversification in the Asia-Pacific region. Food and Agriculture Organization of the United Nations, Bangkok

FAO (2014) Organic agriculture: FAQ. http://www.fao.org/organicag/oa-faq/oa-faq5/en/. Retrieved 5 July 2014
Gauchan D, Smale M, Maxted N, Cole M, Sthapit BR, Jarvis D, Upadhyay MP (2005) Socioeconomic and agroecological determinants of conserving diversity on-farm: the case of rice genetic resources in Nepal. Nepal Agricult Res J 6:89–98
Hoffmann JP (2010) Linear regression analysis: applications and assumptions. Brigham Young University, Utah
Johnston GW, Vaupel S, Kegel FR, Cadet M (1995) Crop and farm diversification provide social benefits. Calif Agric 49(1):10–16
Kabuli A, Phiri M (2007) Farmer perceptions, choice and adoption of soil management technologies in maize-based farming systems of Malawi. Food and Agriculture Organization of the United Nations (FAO), Rome
Matusso J, Mugwe J, Mucheru-Muna M (2012) Potential role of cereal-legume intercropping systems in integrated soil fertility management in smallholder farming systems of sub-Saharan Africa. Third RUFORUM Biennial Meeting. The Regional Universities Forum for Capacity Building in Agriculture (RUFORUM), Entebbe, pp 1815–1843
Pandey S (2013) Factors affecting crop diversity in farmers' fields in Nepal. Renew Agricult Food Syst 30(2):1–8
Rehima M, Belay K, Dawit A, Rashid S (2013) Factors affecting farmers' crops diversification: evidence from SNNPR, Ethiopia. Int J Agricult Sci 3(6):558–565
Scialabba NE-H (2007) Organic agriculture and food security. International Conference on Organic Agriculture and Food Security. Food and Agriculture Organization of the United Nations (FAO), Rome, pp 1–22
Sipiläinen T, Marklund P-O, Huhtala A (2008) Efficiency of agricultural production of biodiversity: Organic vs. conventional practices. 107th EAAE Seminar "Modeling of Agricultural and Rural Development Policies", Sevilla, pp 1–23
UNCTAD (2003) Organic fruit and vegetables from the tropics: market, certification and production information for producers and international trading companies. United Nations Conference on Trade and Development (UNCTAD), Geneva
UNEP-UNCTAD (2008) Organic agriculture and food security in Africa. UNEP-UNCTAD capacity-building task force on trade, environment and development, New York/Geneva
Wilsey BJ, Potvin C (2000) Biodiversity and ecosystem functioning: importance of species evenness in an old field. Ecology 81(4):887–892
Zhang H, John R, Peng Z, Yuan J, Chu C, Du G, Zhou S (2012) The relationship between species richness and evenness in plant communities along a successional gradient: a study from sub-alpine meadows of the eastern Qinghai-Tibetan Plateau, China. PLoS One 7(11):e49024

Chapter 8
Income from Organic and Conventional Farming Systems

Abstract This chapter assesses farm income (total farm valuation) and gross farm cash income (income from selling crops in the market) generated among organic and conventional farmers and analyzes factors impacting them taking into consideration the existence of premium market. Data from 285 respondents from Chitwan District of Nepal, selected using stratified sampling method, were analyzed using bivariate probit and ordinary least square model. Among other socioeconomic variables, this chapter finds that income from organic farming is lower because production per hectare, commercialization rate, and price at which the crops are sold per unit is higher for conventional farming, and access to premium market for organic products is very limited. Linkage with local organic market is necessary, while training, increasing farm size, access to credit, market information, and diversified crops equally play important role to improve income from farming.

8.1 Introduction

Nepal is predominantly an agriculture-based economy that accounts for 36% of gross domestic product and employs 66% of 26.5 million people (MoAD 2015). Therefore, progress in this sector is very much essential for improving lives of the majority and for overall economic development. Monetary benefit is one of the major driving forces for farmers as it provides resources to re/invest in not just farming activities but other sectors too, which ultimately improves their living standard. According to Ramdhani and Santosa (2012), economic justification plays an important role for smallholder farmers than social and environmental benefits to sustain with their farming enterprise in a long run. Especially in developing countries, where smallholder farmers contribute over 80% of food consumed, income still plays a vital role followed by environmental, technological, social, and political aspects (IFAD 2013). Within this sector, there is a growing interest in economic viability of organic farming compared to conventional farming.

Conventional farming is known for its profit orientation. Although massive breakthrough in agricultural technologies pertaining to conventional farming brought remarkable changes in food productivity (IFPRI 2002), it was later criticized for it brought various environmental, economic, and social concerns (DFID

2004; Kassie and Zikhali 2009). Organic farming, on the other hand, is conceived to be one of the most sustainable approaches to food production system, an alternative to ecologically unsound practices of conventional farming. It combines tradition, innovation, and science to adapt to local conditions and sustain the health of soil, ecosystem, and people (IFOAM 2014; IFOAM 2009). Organic farming, though provides social and environmental benefits, the argument over its monetary return is the major bottleneck for its large scale adoption.

In case of organic farming, income may increase through improved yield along with the combination of reduced cost. But it is the premium that attracts farmers to convert which usually makes up for any yield or productivity losses during the transition (Giovannucci 2005). In Nepal, in addition to the export market for organic products (DoAE 2006; Tamang et al. 2011; Pokhrel and Pant 2009), local market in urban areas is also on rise (Willer and Kilcher 2009; Willer and Kilcher 2010). However, marketing is usually done unsystematically on the basis of community trust (Sharma 2005). Without certification, some farmers are able to get premium price based on this mutual trust or personal links, whereas others are devoid of such benefit despite of being certified because of poor marketing system and skill (Singh and Maharjan 2013). Thus, profitability of organic farming through access to premium market cannot simply be explained by the fact that it is certified, especially in the case of local market in Nepal.

The objective of this chapter is to analyze the difference in farm income (total farm valuation) and gross farm cash income (income from selling crops in the market) between organic and conventional farming. Because farm income includes valuation of overall farm output from crops that were sold in the market and those self-consumed as well, it gives us a complete picture of how much organic and conventional farming is contributing to household farm income. As for gross farm cash income (hereafter referred to as "farm cash income"), market involvement of farmers for the purpose of selling crops and an extent to which they are able to generate income thereof were analyzed. It takes into consideration the existence of premium market for organic products, either local or export based. Farm cash income is calculated as the monetary income obtained from selling cereals, vegetables, spices, pulses, oil seeds, and/or fruits in the market without deducting the cost incurred, under organic and conventional farming systems.

Farm households can be observed as an autonomous entity that has capacity to make decision to the best of their interest considering their limited resources. Therefore, the study also assesses various socioeconomic characteristics of farmers to analyze their impact on farm income and farm cash income. It is necessary to recognize those traits so as to intervene in those characteristics that can have significant positive implication on income. Overall, this study is expected to help understand the opportunities and challenges within organic market and assist in developing strategies for making it monetarily attractive for the farmers.

8.2 Insight on Farming-Related Income and Its Influencing Factors

The primary issue of this chapter is to analyze how income from organic farming would compare with conventional farming. Both farm income and farm cash income is regarded to be impacted in a similar way by households' socioeconomic characteristics as it is eventually the income this chapter is focused in assessing. As mentioned above, premium price is the most attractive feature for organic farmers, but from the field survey, it is known that premium market for organic products in the local area is nonexistent. However some farmers are able to export their produces in other cities where it does exist (Table 8.1). Such market is only limited to cereal crops such as rice, maize, wheat, and buckwheat, longer shelf life crop like kidney bean, and nonperishable product like honey. Except for carrot, which is the

Table 8.1 Organic products sold by cooperative in Phoolbari VDC (April–May 2012 to March–April 2013)

SN	Item	Quantity sold (kg)	Price (NRs./kg)	Total production (kg)*	Sold (%)	Regular price (NRs./kg)	Premium (%)
I.	Cereals:						
1	*Chamal* (husked rice)	1850	57	100,866 (unhusked)	6	50	14
2	*Dhan* (unhusked rice)	4000	22			20	10
3	*Makai* (maize)	500	31	19,932	3	18	72
4	*Gahu* (wheat)	1450	30	2440	59	18	67
5	*Fapar* (buckwheat)	1200	60	1595	75	25	140
II.	Pulses:						
6	*Rajma* (kidney bean)	605	120	2053.5	29	70	71
III.	Vegetable:						
7	*Gajar* (carrot)	5000	12	78,407	6	11	9
	Total	14,605	–	205,293.5	7	–	–
IV.	Others:						
8	*Maha* (honey)	121.5	300	(no data)	–	–	–

Source: Field survey (2014)

Note: Total production (kg)*signifies total amount of respective crops produced organically by only those (organic) farmers who are member of a cooperative through which they are sold at premium market in other cities most commercialized non-staple crop mainly in Phoolbari VDC, most vegetables as of present could not be exported due to its easily perishable nature and lack of other facilities to maintain its quality. Currently only 7% of the crops produced organically are sold in the premium market. The rest are sold in local market at the same price as conventional products. With this scenario, it is expected that organic farming could have either higher or lower income compared to conventional farming (Table 8.2).

Socioeconomic variables were chosen and their influence on income is hypothesized based on number of literatures and situation within the local context itself. The definition and measurement of each variable along with their hypothesized relation on income is provided in Table 8.2. Education might have negative impact on farming-based income since educated people switch occupation to be better compensated for their work. On the other hand, it could also have impacted positively on agricultural productivity and indirectly as an external source of

Table 8.2 Definition and measurement of selected variables for farm and cash income

Variables	Definition and measurement	Expected sign	References
Dependent			
Farm_income	Monetary and non-monetary gross value of total production from farming of cereals, vegetables, spices, pulses, oil seeds, fruits, livestock products and by-products, and occasional income generated from selling trees; in Nepalese rupees (NRs)/year	–	–
Cash_income	Income from selling crops; in NRs/ha/year	–	–
Independent			
Farm_method	Farmers practicing organic farming; 1 = yes, 0 otherwise	+ve/-ve	Own elaboration
HHHgender	Male-headed HH; 1 = yes, 0 otherwise	+ve	FAO (2015)
HHHage	Age of HHH; in years	-ve	Own elaboration
HHHedu	Education of HHH; in years	+ve/-ve	Mahmudul et al. (2003) and Weir (1999)
HHHprimary_occu	Primary occupation of HHH; 1 = farming, 0 otherwise	+ve	Own elaboration
Rent	Farmers renting farmland by paying either cash or through crop sharing or mortgaging in farmland; 1 = yes, 0 otherwise	-ve	Own elaboration
Org_exp	Experience of practicing organic farming; in years	+ve	Alexopoulosa et al. (2010)

(continued)

8.2 Insight on Farming-Related Income and Its Influencing Factors

Table 8.2 (continued)

Variables	Definition and measurement	Expected sign	References
LFU	Labor force availability in HH; in labor force unit (LFU)	+ve	Pattanapant and Shivakoti (2009), Gauchan et al. (2005), Adil et al. (2004), and Parvin and Akteruzzaman (2012)
LSU	Livestock holding in HH; in livestock unit (LSU)	+ve	Adil et al. (2004)
Farm_size	Operational farm size; in ha	+ve	Rahman (2010) and Mahmudul et al. (2003)
Non-farm_income	Income from non-farm activities (service, business, rent, remittance, pension, and laboring); in NRs./HH/year	+ve/-ve	Adolwa et al. (2010), Akinola and Owombo (2012), and Gauchan et al. (2005)
Membership	Being in/formal group member formed for organic farming; 1 = yes, 0 otherwise	+ve	Adil et al. (2004)
Org_training	Organic farming-related training; number of times	+ve	Adil et al. (2004)
VDC	Belonging to Phoolbari VDC; 1 = yes, 0 otherwise	+ve	Own elaboration
Agrovet	An exclusive store for agriculture-related products where agricultural inputs (seeds, fertilizers, pesticides, etc.) and equipment and livestock such as veterinary medicine could be found; distance to nearest agrovet (in km)	-ve	Adil et al. (2004)
Market	Market for selling agricultural products; distance to nearest market (in km)	-ve	Adil et al. (2004)
Credit	Credit taken for farming related activities; 1 = yes, 0 otherwise	+ve	Shah et al. (2008)
Final_price	Know price of one or more crops at which it is sold to consumers; 1 = yes, 0 otherwise	+ve	Own elaboration
Commercialization	Commercialization rate (total quantity of crops sold/total produced)	+ve	Own elaboration
SHDI	Shannon Diversity Index (SHDI) that captures both crop richness and evenness	+ve	Padmavathy and Poyyamoli (2012)

income for risk aversion and to overcome credit constraints in farming (Mahmudul et al. 2003; Weir 1999). Similar relation of non-farm income is expected that it could either reduce the significance of having to earn through farming activities (Adolwa et al. 2010) or contribute as a credit relief/financial support for expanding the marketing activities (Akinola and Owombo 2012; Gauchan et al. 2005). Higher labor means higher manpower to contribute in farming activities (Pattanapant and Shivakoti 2009; Gauchan et al. 2005; Adil et al. 2004; Parvin and Akteruzzaman 2012) including production and marketing the crops. Livestock has positive effect too as it improves productivity through supply of manure (Adil et al. 2004). Farm size also has positive relation to farming-based income as people who have more land can produce more crops (Rahman 2010; Mahmudul et al. 2003). Factors such as membership and training are expected to complement the capacity, skill, and information required for improving farming-based income as shown by Adil et al. (2004) that complementary factors like seed, fertilizer, and irrigation cost can have positive effect on income of farmers. In this regard, agrovet and market are also important associations through which farmers can improve their farming performance by having easy access to inputs or getting information on market scenario. Thus, farther these associations are to the farming household, lower the farm-based income is to be expected. Shah et al. (2008) showed that credit facility allowed farmers to allocate more land to different crops compared to non-borrowers, thus leading to increased crop yield and income significantly.

Besides these, other variables considered are gender of HHH, age of HHH, experience of practicing organic farming, primary occupation of HHH, tenancy status, labor availability, HH belonging to Phoolbari VDC, knowing price of crops paid by consumers, commercialization rate, and SHDI. Except for age of HHH and tenant farmers, all are expected to have positive relation to farming-based income. In farming, men are expected to produce more than females as they have better access to resources and information (FAO 2015). With higher experience of organic farming, it is supposed that farmers would have become skilled on various techniques (Alexopoulosa et al. 2010) to improve production and marketing. Farmers who rely on farming as their primary occupation would be more determined to earn higher farming-based income to support the family.

Farmers in Phoolbari VDC are expected to have higher income than those in other two VDCs because at present premium market is mainly accessible to farmers in this VDC. On the other hand, farmers who know the price of crops paid by consumers make informed decision on which crops to produce and market. Although it is understandable that commercialization rate will increase income, this study also assesses its marginal impact that assesses by how much rate increase in commercialization will increase the income. A study by Padmavathy and Poyyamoli (2012) showed that higher diversity will lead to higher gross income, and so higher SHDI is expected to have positive impact on farming-based income. Age of HHH is believed to have negative impact because with age, one's capacity to work declines which could impact on production and marketing activities. Finally, tenant farmers', who are known to be resource poor, ability to invest in production enhancing inputs is perceived to be low. Also because they will have to pay half of their produce as a rent, it is likely that they will be left with very less or no crops for selling in the market.

8.3 Empirical Model

Statistical Software – STATA 13 was used to regress the models for the analysis. For simplicity, all redundant variables that have insignificant P-value were identified through backward elimination method and removed. Thus, it can be said that the empirical specification for all the models provided below is the best subset of predictors. Ordinary least square (OLS) model is used to analyze socioeconomic factors impacting farm income between organic and conventional farming, which can be expressed as:

$$y_i = \beta_0 + x_i\beta_i + \varepsilon_i \tag{8.1}$$

where y_i is farm income, x_i is socioeconomic variables, i is number of observations, β_0 is coefficient of intercept, β_i is parameter to be estimated, and ε_i is error term.

The empirical specification for OLS model can be given by:

$$\begin{aligned}\ln\_\text{farm\_income} =\ & \beta_0 + \beta_1\text{farm\_method} + \beta_2\text{HHHgender} \\ & + \beta_3\text{HHHage} + \beta_4\text{HHHedu} + \beta_5\text{HHHprimary\_occu} \\ & + \beta_6\text{rent} + \beta_7\text{LFU} + \beta_8\text{LSU} + \beta_9\text{farm\_size} \\ & + \beta_{10}\ln\_\text{non-farm\_income} + \beta_{11}\text{org\_training} + \beta_{12}\text{VDC} \\ & + \beta_{13}\text{agrovet} + \beta_{14}\text{market} + \beta_{15}\text{credit} + \beta_{16}\text{SHDI} + \varepsilon\end{aligned} \tag{8.2}$$

where ln is natural log.

Secondly, this study assesses (gross) farm cash income at individual household level. The sample is such that there are number of households who are not engaged in selling their farm products, meaning they utilized their farm produces solely for own household consumption. Table 8.3 shows that 79% of households sell their produce in the market, while 21% of them produce for their own household consumption only. On average, households generated farm cash income worth NRs.107,420 per ha from selling crops in the market.

Although OLS is the most frequently used model for fitting the regression line, it could give biased parameter estimates arising from a missing data problem. The Heckman selection model has been introduced to address this problem of sample selection where only partial observation is made from the outcome variable (Heckman 1979). However, when this model was applied for this data, it gave insignificant (p-value $= 0.482$) lambda value (14865.2). Since lambda is a product of rho and sigma (where rho is correlation between errors in selection and outcome equations and sigma is error from outcome equation), it can be implied that the problem of sample selection bias remains minimal. According to Kennedy (1998),

Table 8.3 Measurement and summary of dependent variables of market involvement and cash income

Dependent variables	Definition and measurement	Bivariate probit model %	OLS model Mean ± SD
Market_involvement	Involved in selling crops in the market; 1 = yes, 0 otherwise	79%	–
Cash_income	Income generated from selling crops in the market (without deducting cost of production); in NRs./ha	–	107419.6 ± 105576.1

Source: Field survey (2013)

trivial correlation between errors of outcome and selection equations is one of the reasons why Heckman model does not perform well. In such case with no selectivity bias, two methods can be analyzed separately (probit for probability of being selected and OLS on non-censored observations). Bivariate probit model (BPM) for assessing socioeconomic variables' impact on farm cash income can be expressed as:

$$y_j^* = \beta_0 + x_j\beta_j + e_j \tag{8.3}$$

$$y_j = \begin{cases} 1 & \text{if } y_j^* > 0 \\ 0 & \text{otherwise} \end{cases} \tag{8.4}$$

where j is number of observations, y^* is unobservable latent variable, y is binary variable of whether a household earns farm cash income from selling crops or not, x is socioeconomic characteristics of a household, β_0 is coefficient of intercept, β_i is parameter to be estimated, and e is normally distributed error term.

Marginal effect for BPM is given by:

$$\frac{\partial P\left(y_j = 1/x_j\right)}{\partial x_j} = \varphi\left(x_j\beta\right)\beta \tag{8.5}$$

where φ is distribution function for standard normal random variable.

The empirical specification for BPM can be given by:

$$\begin{aligned}\text{Market involvement} = \ & \beta_0 + \beta_1 \text{HHHgender} + \beta_2 \text{HHHedu} \\ & + \beta_3 \text{HHHprimary\_occu} + \beta_4 \text{rent} + \beta_5 \text{LFU} + \beta_6 \text{LSU} \\ & + \beta_7 \text{farm\_size} + \beta_8 \text{ membership} + \beta_9 \text{credit} \\ & + \beta_{10} \text{final\_price} + e \end{aligned} \tag{8.6}$$

where ln is natural log.

OLS model for assessing socioeconomic variables' impact on intensity of farm cash income earned by a household can be expressed as:

$$y_k = \beta_0 + x_k \beta_k + \mu_k \tag{8.7}$$

where k is number of observations, y_k is observed values of gross farm cash income, x_k is socioeconomic characteristics of HHs, β_0 is coefficient of intercept, β_k is parameter to be estimated and μ_k is error term.

The empirical specification for OLS model can be given by:

$$\begin{aligned}\text{Gross farm cash income} = &\ \beta_0 + \beta_1 \text{farm\_method} \\ &+ \beta_2 \text{rent} + \beta_3 \text{LFU} + \beta_4 \text{VDC} + \beta_5 \text{market} \\ &+ \beta_6 \text{final\_price} + \beta_7 \text{commercialization} + \mu\end{aligned} \tag{8.8}$$

where gross farm cash income is calculated in NRs./ha and ln is natural log.

As for the diagnostic tests, the variation inflation factor (VIF) gave a value below 10 indicating multicollinearity among the variables regressed with farm income, market involvement, and farm cash income does not exist. Likewise, Breusch-Pagan/Cook-Weisberg showed significant P-value for all there models, thus rejecting null hypothesis of homoscedasticity. This suggests there are linear forms of heteroscedasticity. Again White's test did not show significant P-value for any of the three models, implying that there is no problem of nonlinear forms of heteroscedasticity, i.e., variance of error term is constant. To correct heteroskedasticity of any kind, following Nhemachena and Hassan (2007), model estimation was conducted using robust standard errors that neither changes significance of the model nor the coefficients, but gives relatively accurate P-values and is an effective way of dealing with heteroskedasticity (Wooldridge 2006). Additionally, endogeneity test between farm income and farm size was carried out using Hausman test which showed insignificant residual value. Thus, simple OLS model is used instead of estimating instrumental variables.

8.4 Socioeconomic and Farming System Impact on Farming-Related Income

8.4.1 Farm Income

The P-value for regression of OLS model for farm income is highly significant at 1%, which supports the existence of relationship between independent and dependent variables (Table 8.4). The R^2 value of 39% suggests total variation in value of dependent variable explained by the selected independent variables.

Table 8.4 Result from ordinary least square model for farm income

Variables	Coefficient	P-value
Farm_method	−0.11	0.336
HHHgender	0.09	0.615
HHHage	0.01	0.201
HHHedu	0.02	0.054*
HHHprimary_occu	0.37	0.001***
Rent	0.10	0.327
LFU	−0.004	0.890
LSU	0.12	0.003***
Farm_size	0.74	0.000***
ln_non-farm_income	−0.01	0.235
Org_training	0.03	0.309
VDC	−0.14	0.156
Agrovet	0.01	0.616
Market	0.04	0.005***
Credit	0.21	0.195
SHDI	0.46	0.008***
Constant	9.14	0.000***

Source: Field survey (2013)
Note: ***1%, **5%, and *at 10% level of significance
Number of observations = 285, Prob > F = 0.0000***
$F(16, 268) = 8.43$, $R^2 = 0.3899$, Root MSE = 0.735

Compared to conventional farming, organic farming has lower farm income by 11% despite having access to the premium market. Thus, it can be implied that limited access to premium market is not enough to increase farm income. This lower farm income might also have to do with lower production in organic farming. Male-headed households have 9% higher farm income than their female counterparts. It complies with study by FAO (2015) that men are expected to produce more than females not because they are more efficient but because females are often deprived of accessing productive resources and opportunities. Compared to males, they control less land, use fewer inputs, and have less access to important services such as extension advice. This can very much be applied to Nepalese society as it is mainly patriarchal-based. A year increase in age of HHH will actually increase farm income by 1%, which is opposite to the previous hypothesis. According to Alexopoulosa et al. (2010), with age comes valuable experience and expertise which might have led them to be more efficient, thus leading to higher farm income. A year increase in formal education will increase farm income by 2%, significant at 10%, because with education comes one's ability to take better decision or could impact indirectly by channeling income from high remunerative job toward farming activities, as mentioned by Mahmudul et al. (2003) and Weir (1999). Farmers whose primary occupation is farming have farm income higher by 37% than those whose primary occupation is non-farm based, significant at 1%, which

8.4 Socioeconomic and Farming System Impact on Farming-Related Income

indicates that such farmers would be more determined to improve their farm income to support their family.

Surprisingly those who have rented in farmland in addition to having their own farmland will have farm income higher by 10%, compared to those who fully own the farmland. This shows that tenant farmers are more driven toward increasing their farm income as they will have to pay rent in kind while be self-sufficient in household food consumption as well. A unit increase in LFU will decrease farm income by 0.4%. In the study areas, there is a growing tendency of shift in labor toward non-farm sector. This also complements the earlier explanation of male-headed households being attracted toward non-farm sector. In addition to men, younger generation also act less as a helping hand for farming-related activities and are rather engaged in studying to prepare themselves for occupation other than farming in the future. Thus, unlike the conservative belief as has been mentioned in various literatures such as Pattanapant and Shivakoti (2009), Gauchan et al. (2005), Adil et al. (2004), and Parvin and Akteruzzaman (2012), labor is not a defining factor anymore for explaining vibrancy in farming activities and hence the farm-based income. A unit increase in LSU increases farm income by 12%, significant at 1%. Higher livestock holding means higher manure availability that improves production (Adil et al. 2004) which ultimately increases the farm income. A hectare increase in farm size increases farm income by 74%, significant at 1%. Bigger farm size means larger area to accommodate more crops (Rahman 2010; Mahmudul et al. 2003). A percent increase in non-farm income will decrease farm income by 1%, suggesting that non-farm income is being utilized in sectors other than farming as has been found by Adolwa et al. (2010) as well.

As suggested by Adil et al. (2004), training complements the knowledge required to improve farm income, and thus one more training related to organic farming will actually lead to 3% increase in farm income. Farmers in Phoolbari VDC have farm income less by 14% compared to those in other two VDCs. There needs to be more in-depth study to identify what factors might have contributed to having lower farm income. A kilometer distance to agrovet and market increases farm income by 1% and 4%, respectively, the latter significant at 1%. This suggests that the opportunity cost of commuting longer distance to such establishments to buy related inputs/crops is high enough to make farmers cultivate varieties of high-value crops prioritizing self-consumption, thus increasing the overall farm value. Farm income of those who has access to credit is higher by 21% than those who do not. As indicated by Shah et al. (2008), credit accessibility helps improve farm productivity. In this case, farmers used credit for investing in commercial crops, in livestock rearing, and/or for irrigation purposes. SHDI captures both richness (number of species cultivated) and evenness (even distribution of species) of species diversity. Wilsey and Potvin (2000) found that species evenness has more linear relationship with total productivity than with species richness. SHDI gives a better understanding of the status of diversity, and a unit increase in SHDI increasing farm income by 46%, significant at 1%, suggests that it ultimately leads to improved productivity, thus increasing the farm income.

Table 8.5 Result from bivariate probit (selection) model for marketing crops

Variables	Coefficient	Marginal effect	P-value
HHHgender	−0.40	−0.08	0.150
HHHedu	0.04	0.01	0.078*
HHHprimary_occu	0.47	0.11	0.041**
Rent	−0.55	−0.15	0.021**
LFU	−0.1	−0.02	0.051*
LSU	−0.04	−0.01	0.430
Farm_size	1.43	0.33	0.001***
Membership	−0.37	−0.08	0.047**
Credit	0.65	0.11	0.015**
Final_price	0.91	0.16	0.000***
Constant	0.7		

Source: Field survey (2013)
Note: ***1%, **5%, and * at 10% level of significance
Number of observations = 285, Wald chi^2 (10) = 35.97
Log pseudo likelihood = −118.68129, Prob > chi^2 = 0.0001***, Pseudo R^2 = 0.1909

8.4.2 Market Involvement

The *P*-value for regression of BPM model for market involvement is highly significant at 1%, which supports the existence of relationship between independent and dependent variables (Table 8.5). The Pseudo R^2 value of 19% suggests total variation in value of dependent variable explained by the selected independent variables.

Unlike farm income, male-headed households' probability of marketing crops decreases by 8%. In this study area, it is especially men who have the tendency of being engaged in non-farm activities such as teaching, carpentry, etc. within the local area or migrate in other cities or countries for better opportunities. Because of diversified income source, they are less interested in generating income from selling crops, but effort to improve overall farm income still remains intact. A year increase in formal education increases the probability of selling crops by 1%, significant at 10%, which further supports the argument given for farm income that education leads to better decision-making or could impact indirectly by channeling income from high remunerative job toward farming activities (Mahmudul et al. 2003; Weir 1999). Farmers whose primary occupation is farming have higher likelihood of selling crops in the market by 11%, significant at 5%. Although those who have rented in farmland in addition to having their own farmland might have made them competitive to increase farm income, it does not necessarily indicate they are financially well off. Their probability of marketing crops decreases by 15%, significant at 5%, compared to those who fully own the farmland. A unit increase in LFU will contribute negatively to the probability of selling crops by 2%, significant at 10%. Like earlier explanation, because of growing tendency to work in non-farm sector, higher labor does not seem to be translated into higher

8.4 Socioeconomic and Farming System Impact on Farming-Related Income

Table 8.6 Comparing farm cash income across two farming systems from six crops that were partly sold in the premium market

Crops	Farming systems (Mean ± SD)						
	Organic	n^a	Conventional	n	Total	N^b	T-test
Sold_rice	24591.36± 21275.01	44	34710.23± 35674.07	87	31311.53± 31853.26	131	0.086*
Sold_maize	7781.053± 8552.007	19	7932.456± 11844.1	57	7894.605± 11,059	76	0.959
Sold_wheat	7180± 10346.36	8	6125.455± 4232.738	33	6331.22± 5765.859	41	0.648
Sold_buckwheat	2962.909± 3081.677	11	4000± 3605.551	3	3185.143± 3082.239	14	0.625
Sold_kidneybean	10360.29± 11249.72	17	19319.74± 52738.88	38	16550.45± 44280.07	55	0.493
Sold_carrot	41659.09± 48677.1	11	93824.35± 139994.8	31	80162.02± 124326.9	42	0.236

Source: Field survey (2013)
Note: * at 10% level of significance; n^a stands for number of farmers in the respective farming system, and N^b stands for total number of farmers

involvement in farming activities, including selling farm products. A unit increase in LSU contributes negatively to the probability of selling crops by 1%. Livestock rearing is a time-consuming task, which might have led to less time available for marketing the crops. A hectare increase in farm size increases prospect of marketing crops by 33%, significant at 1%. Bigger farm size means larger area to accommodate more crops (Rahman 2010; Mahmudul et al. 2003), the excess of which after household consumption could be sold in the market.

Group membership, conversely, reduces likelihood of marketing the crops by 8%, significant at 5%, despite of it being a platform for organic farmers through which they could sell their crops in the premium market. As shown in Table 8.1, demand for both crop varieties and quantities to be sold in premium market itself is very limited that when income from organic farming is compared with conventional farming, it does not show any significant difference (Table 8.6).

This was further confirmed during focal group discussion where member farmers complained of limited demand not justifying the transportation and opportunity cost of delivering their products to the collection center. The rest are sold in local market at same price as conventional products in which case organic products are often at the losing end as some farmers claim that organic products are visually unattractive. Incidences of vegetables being disfigured, dull colored with less brightness/shine (Fig. 8.1), smaller in size and/or even having small holes from pest attack were often provided; although some said that with enough fertilizer, crop size can be improved and some had the opinion that smaller size might be the result of growing vegetables during off-season.

Although member farmers are given marketing-related training, it is mainly confined to basics such as presentation of organic agro-products for visual attraction, informing consumers of health benefits of consuming organic, and information

Fig. 8.1 Visual difference in organic (left) and conventional (right) brinjal found in the local market (*Source*: Field survey 2013)

of few premium markets in other cities which is outside their jurisdiction. What little access to premium market, which was realized after years of associating with various stakeholders, also comes with its own set of challenges. The previous experiences of dealers mixing their products with conventional ones so that more quantities could be sold at the premium market as organic have led farmers to limit with few trustworthy dealers. Even then they face the problem of late payment through such marketing channel which is worrisome as farmers have to rely on uninterrupted income from one season for an investment in another. Moreover, they also had an unpleasant experience of losing from mass cultivation of *tulsi* (*Ocimum tenuiflorum L.*), which they produced with the intention of exporting it through a private organization. But since the market price considerably plummeted, the whole transaction got discarded, leaving most of the farmers with no option but to utilize the excess crop as green manure. All this combination of incidents made farmers more hesitant toward participating in the premium market and rather sell their produce at the local market even though it means "no premium price." In the local area, very recently there has been influx of shops claiming to sell organic or eco-friendly agro-products, but so far majority of farmers are unaware of it.

Access to credit also increases probability of marketing crops by 11%, significant at 5%. As implied by Shah et al. (2008), credit facility will let farmers have access to means that improves their market participation. Those farmers who are informed of final price at which their products are sold to the consumers show increased probability of selling crops by 16%, significant at 1%, which signifies the importance of market-based information.

8.4.3 Farm Cash Income

The P-value for regression of OLS model for farm cash income is highly significant at 1%, which supports the existence of relationship between independent and

8.4 Socioeconomic and Farming System Impact on Farming-Related Income

Table 8.7 Result from bivariate probit model for marketing crops and ordinary least square model for gross farm cash income

Variables	Coefficient	P-value
Farm_method	−15,408	0.010***
Rent	−9042	0.153
LFU	−1198	0.345
VDC	10,158	0.082*
Market	−795	0.332
Final_price	12,579	0.082*
Commercialization	55,773	0.000***
Constant	30,270	0.001***

Source: Field survey (2013)
Note: ***1%, **5%, and *at 10% level of significance
No. of observation = 225, Prob > $F = 0.0000$***, $R^2 = 0.4941$
$F(7, 217) = 24.17$, Root MSE = 40,606

dependent variables (Table 8.7). The R^2 value of 49% suggests total variation in value of dependent variable explained by the selected independent variables.

Organic farmers tend to earn NRs.15,408/ha less farm cash income than conventional farmers, significant at 1%. Further analysis showed that organic farmers have lesser production from all crop categories except fruits (Fig. 8.2), lower commercialization rate (Fig. 8.3), and lower price per unit received except for cereals (Fig. 8.4). As mentioned above, though farmers could get premium ranging from 9 to 140%, only 7% of crops produced by member organic farmers are currently traded in such market. This shows that access to premium market is not able to contribute significantly in the overall income. Thus, in addition to what Trewavas (2002) and Meisner (2007) stated that organic production is generally lower, the lower commercialization rate and price also contributed equally to their lower farm cash income.

Those renting farmland in addition to having their own farmland have lower probability of generating farm cash income by NRs.9042/ha, compared to those who fully own the farmland. As per our earlier expectation, partial tenant farmers though have higher farm income, it does not necessarily indicate they are financially well off. Because most of them are paying rent through crops, their probability of having excess crops to sell decreases and so is their capacity to earn higher farm cash income by selling crops. A unit increase in LFU will decrease farm cash income by NRs.1198/ha. This is consistent with negative relation of LFU to farm income and being involved in marketing crops. Farming is no longer appealing for larger portion of the farmers as they are diverting more toward non-farm sector.

Farmers in Phoolbari VDC have farm cash income higher by NRs.10,158/ha, significant at 10%, compared to those in other two VDCs. This is in contrast to farmers in Phoolbari VDC having lower farm income. While more in-depth study needs to be done to understand this issue fully, it is at least clear that other two VDCs have fairly lower farm cash income. A kilometer distance to market decreases farm cash income by NRs.795/ha. This also complies with the previous assertion that longer distance to market would discourage farmers from marketing

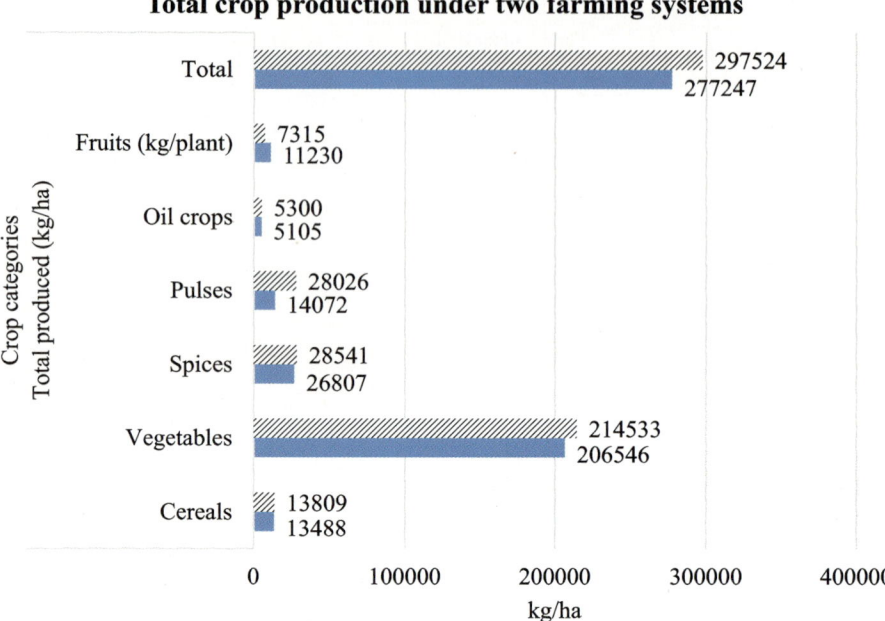

Fig. 8.2 Total crops produced (kg/ha) under two farming systems (*Source*: Field survey 2013)

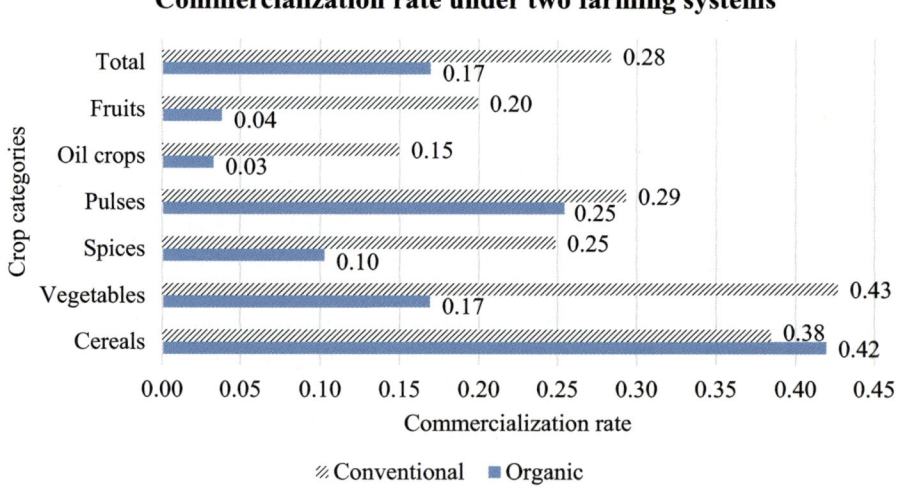

Fig. 8.3 Commercialization rate of two farming systems (*Source:* Field survey 2013)

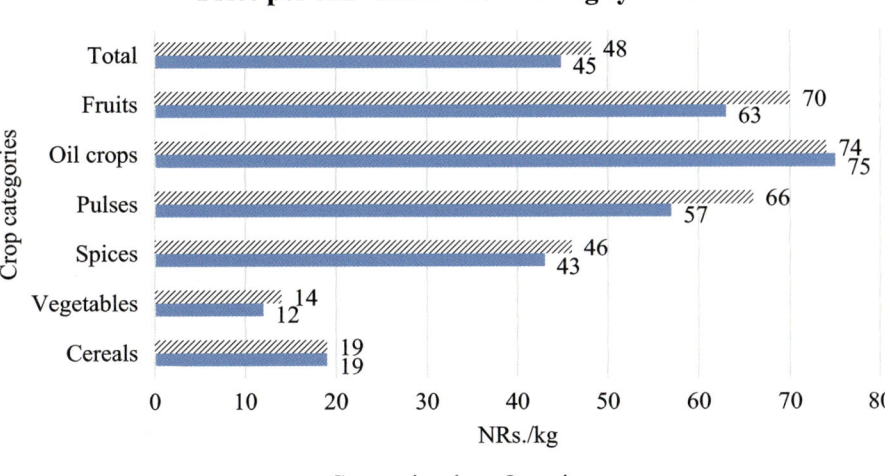

Fig. 8.4 Price per unit of crop under two farming systems (*Source*: Field survey 2013)

their crops as it would take considerable time and cost while commuting. Thus, farmers would rather produce for own household consumption rather than sell or even having to buy from distant market. Those farmers who are informed of final price at which their products are sold to the consumers are expected to have farm cash income higher by NRs.12,579/ha, significant at 10%, which again signifies the importance of market-based information. It is a well-known fact that commercialization of farming is pursued for generating higher income. In this case, a unit increase in commercialization rate increases farm cash income by NRs.55,773/ha, significant at 1%.

8.5 Summary

This chapter tried to analyze income from farming among organic and conventional farmers in an effort to enhance monetary benefit from farming activities, especially from organic farming. Further the income is divided into total farming output valuation and farm cash income generated from selling in the market. The rationality for analyzing these two issues separately against selected set of farmers' socioeconomic characteristics was justifiable from the result which shows some variables having differing impact. Male-headed households, for example, increase farm income, but their inclination to market crops is negative. The growing tendency of shift among males and young generation alike in non-farm sector has put a strain on farming activities. This can further be validated from labor availability in the household which has negative impact on both farm income, being engaged in marketing of crops and generating higher farm cash income therein. In addition to that, non-farm income is also least likely to be invested in farming activities as its impact on farm income is also negative. Likewise, tenant farmers are encouraged to increase farm income but have lower potential of selling crops in the market and generating farm cash income therein because of having to pay crops produced as rent. Livestock holding also increases farm income, but it does not translate into being involved in marketing with an equal intensity as livestock rearing is time-consuming, leaving households with less time for marketing purposes.

Household head's age plays a crucial role in increasing farm income while their education, farming as their primary occupation, and access to credit are important for increasing both farm income and increasing their chances of being engaged in marketing the crops. Having more diversified crops with better evenness increases farm income through improved biological activities. Farm size, which reflects the amount of crops being grown, improves both farm income and likelihood of marketing the crops. Market access, credit facility, and market information definitely have important roles to play in marketing the crops as well.

One of the interesting findings from this study is that training related to organic farming has positive impact on farm income, but membership in a group formed for the purpose of organic farming has negative impact on probability of being engaged in marketing the crops. More so, farmers from Phoolbari VDC, where access to premium market exists, have lower farm income but higher farm cash income than farmers from other VDCs. While the latter needs more scrutiny, from this study, it is recommended that providing training that compensates for lack of education, increasing farm size through merging and collaboration, providing access to credit, disseminating market information such as crops' price at which consumers buy, and encouraging more diversified farm with more varieties and evenness should be encouraged to improve income from farming.

Conventional farmers still have higher farm income and earn higher farm cash income than organic farmers because, at present, the production per hectare, commercialization rate, and price per unit for almost all crops are higher for conventional crops. In addition to that, access to premium market is very limited and has not been able to make any significant contribution to organic farmers' income. Since monetary benefit can attract farmers to divert their labor force in farming activities and specifically to boost income pertaining to organic farming, making access to premium market is very imperative. Organic farmers should be linked with potential sellers not just in other cities but within the local area as well where some shops have recently started to sell organic or eco-friendly agro-products so that the farmers would have more control over the price and quality check of their products.

References

Adil SA, Badar H, Sher T (2004) Factors affecting gross income of small farmers in district Jhang-Pakistan. Pak J Life Soc Sci 2(2):153–155

Adolwa IS, Esilaba AO, Okoth P, Mulwa MR (2010) Factors influencing uptake of integrated soil fertility management knowledge among smallholder farmers in Western Kenya. 12th KARI Biennial Scientific Conference: Transforming agriculture for improved livelihoods through agricultural product value chains. Africa Soil Information Service (AfSIS), Nairobi, pp 1146–1152

Akinola A, Owombo P (2012) Economic analysis of adoption of mulching technology in yam production in Osun state, Nigeria. Int J Agricult Forest 2(1):1–6

Alexopoulosa G, Koutsouris A, Tzouramani I (2010) Should I stay or should I go? Factors affecting farmers' decision to convert to organic farming or to abandon it. 9th European IFSA symposium, Vienna

DFID (2004) Agricultural sustainability. Department for International Development, London

DoAE (2006) Proceedings of a first national workshop on organic farming. Directorate of Agriculture Extension (DoAE), Lalitpur

FAO (2015) Men and women in agriculture: Closing the gap. http://www.fao.org/sofa/gender/key-facts/en/. Retrieved 27 July 2015

Gauchan D, Smale M, Maxted N, Cole M, Sthapit BR, Jarvis D, Upadhyay MP (2005) Socioeconomic and agroecological determinants of conserving diversity nn-farm: the case of rice genetic resources in Nepal. Nepal Agricult Res J 6:89–98

Giovannucci D (2005) Organic agriculture and poverty reduction in Asia: China and India focus. International Fund for Agricultural Development (IFAD), Rome

Heckman JJ (1979) Sample selection bias as a specification error. Econometrica 47(1):153–162

IFAD (2013) Smallholders, food security, and the environment. International Fund for Agricultural Development (IFAD), Rome

IFOAM (2009) High sequestration, low emission, food secure farming. IFOAM EU Group, Bonn

IFOAM (2014) Definition of organic agriculture. http://www.ifoam.org/en/organic-landmarks/definition-organic-agriculture. Retrieved 3 July 2014

IFPRI (2002) Green revolution: curse or blessing? International Food Policy Research Institute, Washington, DC

Kassie M, Zikhali P (2009) Sustainable land management and agricultural practices in Africa: bridging the gap between research and farmers. University of Gothenburg, Gothenburg

Kennedy P (1998) A guide to econometrics. The MIT Press, Cambridge

Mahmudul HA, Ishida A, Taniguchi K (2003) The role of farmers' education on income in Bangladesh. Bulletin of Education and Research Center for Lifelong Learning, pp 29–35

Meisner C (2007) Why organic food can't feed the world?. http://www.cosmosmagazine.com/features/online/1601/why-organic-food-cant-feed-world. Retrieved 29 Oct 2011

MoAD (2015) Welcome to Ministry of Agricultural Development (MoAD). http://www.moad.gov.np/. Retrieved 13 March 2015

Nhemachena C, Hassan R (2007) Micro-level analysis of farmers' adaptation to climate change in Southern Africa. Int Food Policy Res Inst (IFPRI), Washington

Padmavathy A, Poyyamoli G (2012) Provisioning ecosystem services income extend comparison between organic and conventional agricultural fields in Puducherry-India. J Agricult Extens Rural Develop 4(6):120–128

Parvin M, Akteruzzaman M (2012) Factors affecting farm and non-farm income of Haor inhabitants of Bangladesh. Progress Agric 23(1–2):143–150

Pattanapant A, Shivakoti GP (2009) Opportunities and constraints of organic agriculture in Chiang Mai Province, Thailand. Asia-Pac Develop J 16(1):115–147

Pokhrel DM, Pant KP (2009) Perspectives of organic agriculture and policy concerns in Nepal. J Agricult Environ 10:89–99

Rahman M (2010) Socio-economic determinants of off-farm activity participation in Bangladesh. Russ J Agricult Socio-Econom Sci 1(13):3–7

Ramdhani MA, Santosa E (2012) Key success factors for organic farming development. Int J Basic Appl Sci 1(1):7–13

Shah MK, Khan H, Jehanzeb KZ (2008) Impact of agricultural credit on farm productivity and income of farmers in mountainous agriculture in northern Pakistan: a case study of selected villages in district Chitral. Sarhad J Agricult 24(4):713–718

Sharma G (2005) Organic agriculture in Nepal: an analysis into status, policy, technology and psychology. National workshop on organic agriculture and food security, Kathmandu

Singh M, Maharjan KL (2013) Prospect of farmers in generating additional income through organic vegetable farming: a case study in Kathmandu valley and Chitwan district of Nepal. J Int Develop Cooperat 19(4):37–49

Tamang S, Dhital M, Acharya U (2011) Status and scope of organic agriculture in Nepal. Food and sustainable agriculture initiative. Forestaction, Lalitpur

Trewavas A (2002) Malthus foiled again and again. Nature 418:668–670

Weir S (1999) The effects of education on farmer productivity in rural Ethiopia - working paper CSAE WPS99-7. Centre for the Study of African economies. University of Oxford, Oxford

Willer H, Kilcher L (2009) The world of organic agriculture: statistics and emerging trends. International Federation of Organic Agriculture Movements (IFOAM)/Research Institute of Organic Agriculture (FiBL), Bonn/Frick

Willer H, Kilcher L (2010) The world of organic agriculture: Statistics and emerging trends. International Federation of Organic Agriculture Movements (IFOAM)/Research Institute of Organic Agriculture (FiBL), Bonn/Frick

Wilsey BJ, Potvin C (2000) Biodiversity and ecosystem functioning: importance of species evenness in an old field. Ecology 81(4):887–892

Wooldridge JM (2006) Introductory econometrics: a modern approach. Thomson South Western, Ohio

Chapter 9
Crop Production and Net Return from Organic and Conventional Farming Systems

Abstract Economic return from organic farming is a highly debated issue. This study was conducted in Chitwan District of Nepal and analyzes three crops, carrot, potato, and cauliflower, that are among the most commercial non-staple crops and form an important part of daily food consumption. Respondents were selected based on stratified random sampling, and cost component of production factors, production, and net return were analyzed comparing organic and conventional farming systems. T-test was used for assessing production factors and net return and ordinary least square model for analyzing farm households' socioeconomic variables' impact on crop production, respectively. Result finds that net return from conventional potato is significantly higher, but overall production of the three crops combined is higher in organic farming system. Male-headed households have positive impact on production, while households having bigger farm size and with farther distance to agrovet have negative impact on production. Among the production factors, land area under cultivation, seed, organic inputs, chemical inputs, and tillage have positive impact. Since chemical inputs can deplete soil fertility over time, crop production should be improved by adopting a more sustainable way, such as boosting use of organic inputs.

9.1 Introduction

Nepal is one of the least developed countries located within South Asia where agriculture is the mainstay contributing 36% to gross domestic product (GDP) and forming source of income and employment for 66% of the population (SECARD 2011; MoAD 2015; MoE 2011). Commercializing farming has been deemed necessary to bring much needed changes in its economic growth (Samriddhi 2011), but its long-term impact through declining soil fertility, negative repercussion on environment and health of farmers due to use of agrochemicals, and market demand helped enforce organic movement in Nepal (Bhatta and Doppler 2010; Weiss 2004). Organic farming is known for being a sustainable food production method that relies on agroecological principles resulting in improvement of soil fertility rather than depleting it. With intensifying issue of climate change, contribution of organic farming to mitigate and adapt to the changes has further boosted its potential (Scialabba 2007). It also ensures a safe working environment by avoidance of use and hence exposure to harmful chemicals and provides residue-free products to the consumers (ESCAP 2012; Vaarst 2010). Although organic farming provides social and environmental benefits, its economic benefit is often debatable when compared to conventional farming. According to Ramdhani and Santosa (2012), economic justification plays an important role for smallholder farmers than social and environmental benefits to sustain with their farming enterprise in a long run. Especially in developing countries, income still plays a vital role followed by environmental, technological, social, and political aspects (IFAD 2013). Therefore, to increase the share of organic farming, which as of 2014 comprises of only 1% of the overall agricultural land (including in-conversion areas) in the global context and only 0.2% in Nepalese context (Willer and Lernoud 2016), ensuring economic benefit is very much essential.

There are various theories to justify yield and overall economic benefit of organic and conventional farming systems. Evolution of farming through the use of modern inputs has been successful in increasing production in various parts of the world. But the use of costly agrochemical inputs has put strain on overall return or its actual economic benefit (FAO 1996; Hazell and Wood 2008). Organic farming is a cost-effective and affordable farming system that does not require expensive technical investment but rather relies on locally available resources (Leu 2011). Contrastingly, it is also suggested that increasing cultivated area under organic farming is inevitable for higher production. Another challenge is to supply enough organically acceptable fertilizer which is difficult to acquire (Trewavas 2002; Meisner 2007). The most important aspect of organic farming is it can generate premium price in an established market that ultimately generates more profit. Therefore, often times, income under organic farming system is better, exclusively or in combination with improved yields, reduced costs, or premium price which can compensate for any yield losses that may incur during the transition phase (Giovannucci 2005). This study is conducted to compare economic benefit from organic farming with respect to conventional farming. The study compares cost components of production factors and analyzes net return from organic and conventional carrot, potato, and cauliflower, which are among the most commercially cultivated non-staple crops and are important part of daily food consumption in the study areas, as has been identified through key informant interview. In fact, these are among the important crops across the country as well. The production of carrot was 26,296 mt produced in 2488 ha, potato was 2,690,421 mt produced in 64,483 ha and cauliflower was 524,205 mt produced in 34,065 ha. Cauliflower is the top most produced vegetable, while carrot is the seventh most produced vegetable in Nepal. Likewise, potato is the second highest cash crop produced after sugarcane in Nepal (MoAD 2013). This chapter also looks into the impact of various socioeconomic factors on crop production, which are as important as agroecological variables and farmers' perception (Kafle 2010).

9.2 Sample Selection and Data Collection

Among 285 respondents chosen randomly by stratifying members and nonmembers of a group formed for the purpose of organic farming in three VDCs, Phoolbari, Mangalpur, and Shivanagar, only those respondents who have grown the respective crops were chosen. Sample size for analysis of organic and conventional carrot, potato, and cauliflower growers based on their VDC and membership status is shown in Tables 9.1 and 9.2, respectively. Data is generated based on recollection of farmers on the respective crops' cultivation of the past 1 year.

9.3 Influencing Factors on Crop Production

Table 9.3 briefs definition, measurement, and descriptive analysis of selected socioeconomic variables, and Table 9.4 provides their expected relation to the crop production. As explained before, production under organic farming is known to be less compared to conventional farming unless land area under organic cultivation can be increased. On the other hand, it is difficult to access enough inputs that are acceptable under standard organic farming (Trewavas 2002; Meisner 2007). Contrastingly some studies have shown organic production to result in yield equal to or higher than that of conventional if practiced in an effective way. A study done on organic farming and the global food supply showed that organic production method has capacity to sustain current or even larger human population without increasing agricultural land base, emphasizing the importance of leguminous cover crops for effective results (Badgley 2007). Thus, it cannot be generalized whether organic farming compared to conventional farming will result in higher or lower production. The feasibility of any farming system very much depends on local characteristics.

In farming, males are expected to produce more than females not because they are more efficient, but because females are often deprived of accessing productive resources and opportunities. Compared to males, they control less land, use fewer inputs, and have less access to important services such as extension advice (FAO 2015). With age, one's ability to garner physical strength diminishes, thus eventually decreasing the farming production. But on the other hand, they are experienced on utilizing labor more effectively through efficient combinations of input (Guo et al. 2015). Thus, direction of this variable on crop production could go either way. Education is one of the most important factors contributing in the development of farming sector. Educated farmers have better access to information and utilize it for producing better results.

Livestock holding and farm size are indicators of wealth status in rural areas and thus is expected to contribute positively to farming production. Livestock provides draft power, manure fertilizer, and farm cash income, while bigger farm size allows opportunity to adopt improved technologies, thus increasing the farming

Table 9.1 Sample distribution across crops based on VDC and farming system

Crops	Sample size		VDC	Organic	Conventional	Total	P-value
	Organic	Conventional					
Carrot	45	36	Phoolbari	22 (48.89)	22 (61.11)	44 (54.32)	0.273
			Others	23 (51.11)	14 (38.89)	37 (45.68)	
Potato	64	55	Phoolbari	35 (54.69)	13 (23.64)	48 (40.34)	0.001***
			Others	29 (45.31)	42 (76.36)	71 (59.66)	
Cauliflower	84	24	Phoolbari	40 (47.62)	3 (12.50)	43 (39.81)	0.002***
			Others	44 (52.38)	21 (87.50)	65 (60.19)	
			No	35 (41.67)	21 (87.50)	56 (51.85)	

Source: Field survey (2013)
Note: ***1% level of significance

9.3 Influencing Factors on Crop Production

Table 9.2 Sample distribution across crops based on membership and farming system

Crops	Sample size		Membership	Organic	Conventional	Total	P-value
	Organic	Conventional					
Carrot	45	36	Yes	26 (57.78)	16 (44.44)	42 (51.85)	0.233
			No	19 (42.22)	20 (55.56)	39 (48.15)	
Potato	64	55	Yes	34 (53.13)	11 (20.00)	45 (37.82)	0.000***
			No	30 (46.88)	44 (80.00)	74 (62.18)	
Cauliflower	84	24	Yes	49 (58.33)	3 (12.50)	52 (48.15)	0.000***
			No	35 (41.67)	21 (87.50)	56 (51.85)	

Source: Field survey (2013)
Note: ***1% level of significance

Table 9.3 Definition and measurement of selected variables for crop production

Variables	Definition and measurement	Carrot Mean ± SD/%	Potato Mean ± SD/%	Cauliflower Mean ± SD/%
Dependent				
Production	Crop production in a year; in kg/ha	18,195 ± 5090	10,614 ± 4487	16,798 ± 16,399
Independent				
Farm_method	Farmers practicing organic farming; 1 = yes, 0 otherwise	33%	54%	78%
HHHgender	Male-headed HH; 1 = yes, 0 otherwise	91%	92%	97%
HHHage	Age of HHH; in years	50.03 ± 11.31	51.24 ± 11.48	49.93 ± 11.92
HHHedu	Education of HHH; in years	7.62 ± 5.40	5.45 ± 5.32	7.07 ± 5.64
HHHprimary_occu	Primary occupation of HHH; 1 = farming, 0 otherwise	56%	64%	50%
Org_exp	Experience of practicing organic farming; in years	3.49 ± 7.75	2.45 ± 6.42	3.32 ± 7.65
LFU	Labor force availability in HH; in labor force unit (LFU)	4.27 ± 1.62	4.27 ± 1.76	4.29 ± 1.79
LSU	Livestock holding in HH; in livestock unit (LSU)	2 ± 1.30	1.82 ± 1.74	1.75 ± 1.38
Farm_size	Operational farm size; in ha	0.50 ± 0.38	0.54 ± 0.41	0.49 ± 0.43
Nonfarm_income	Income from non-farm activities (service, business, rent, remittance, pension, and laboring); in log NRs./HH/year	9.26 ± 5.36	9.58 ± 4.96	10.19 ± 4.67
Membership	Being in/formal group member formed for organic farming; 1 = yes, 0 otherwise	52%	38%	48%

(continued)

Table 9.3 (continued)

Variables	Definition and measurement	Carrot Mean ± SD/%	Potato Mean ± SD/%	Cauliflower Mean ± SD/%
Org_training	Organic farming-related training; number of times	1.59 ± 2.66	0.85 ± 1.55	1.17 ± 1.73
VDC	Belonging to Phoolbari VDC; 1 = yes, 0 otherwise	54%	40%	40%
Agrovet	Store for agriproducts (such as seeds, fertilizers, pesticides, agri-equipment, veterinary medicine, etc.); distance to nearest agrovet (in km)	1.63 ± 1.40	1.5 ± 1.73	1.61 ± 1.80
Market	Market for selling agricultural products; distance to nearest market (in km)	3.30 ± 3.60	2.30 ± 2.98	2.90 ± 3.67
Credit	Credit taken for farming-related activities; 1 = yes, 0 otherwise	10%	8%	10%

Source: Field survey (2013)

production. Off-farm income and credit also contribute in the increase of production as farmers will be able to finance purchase of required inputs and technologies (Tesema 2006). Group membership is expected to increase crop production because such common interest groups have increased chances to access new technologies and have multiplier effect (Akal 2014). Training related to organic farming mainly provided through such group is expected to have positive impact as well since training increases knowledge, skills, and capabilities of farmers to produce in a more conducive way (Ahmad et al. 2007). In the same line, the farther the distance to agrovet and market, it is expected to decrease production because these institutions are the source of information and offer input which could be advantageous in improving crop production.

Labor availability can contribute in intensifying farming activities and hence increases farming productivity and income (Adil et al. 2004; Parvin and Akteruzzaman 2012). It is expected that when farming is HHH's primary occupation, they would be more determined toward increasing production. The longer the experience of organic farming, the higher the production is to be expected because

Table 9.4 Hypothesized relation of independent variables to dependent variable of crop production

Independent variables	Expected sign	References
Farm_method	+ve/-ve	Trewavas (2002), Meisner (2007) and Badgley et al. (2007)
HHHgender	+ve	FAO (2015)
HHHage	+ve/-ve	Guo et al. (2015)
HHHedu	+ve	Tesema (2006)
HHHprimary_occu	+ve	Own elaboration
Org_exp	+ve	Own elaboration
LFU	+ve	Adil et al. (2004) and Parvin and Akteruzzaman (2012)
LSU	+ve	Tesema (2006)
Farm_size	+ve	Tesema (2006)
Non-farm_income	+ve	Tesema (2006)
Membership	+ve	Akal (2014)
Org_training	+ve	Ahmad et al. (2007)
VDC	+ve/-ve	Own elaboration
Agrovet	-ve	Own elaboration
Market	-ve	Own elaboration
Credit	+ve	Tesema (2006)

such experience might have resulted in gaining various skills that could be conducive in improving production level. Lastly, Phoolbari VDC is expected to have higher production as carrot is cultivated distinctively more commercially in this VDC (Figs. 9.1 and 9.2) compared to the other two VDCs, but for other two crops, it could have higher or lower production.

9.4 Empirical Model

All cost components for the respective crops are calculated for each respondent (except for irrigation cost[1]). Local measurements were converted into a standard unit (i.e., NRs. per ha). Cost of production was then calculated using the following equation:

---

[1]Irrigation cost is excluded from the calculation of cost of production because of nonuniformity of data among the respondents. It might have to do with the fact that farmers rely on rainwater, especially if it is not grown on commercial scale. Sometimes water is supplied from household wastewater, tap, or well, which is why they do not have proper accounting of irrigation cost.

Fig. 9.1 Commercial cultivation of carrot in Phoolbari VDC (*Source*: Field visit 2015)

Fig. 9.2 Carrots being washed on a commercial scale (*Source*: Field visit 2015)

$$Cp = Cland + Cseed + Corg + Clitter + Cinorg + Ctillage + Cwage + Cpp \qquad (9.1)$$

where Cp = total expenditure; Cland = cost of land; Cseed = cost of seed; Corg = cost of organic fertilizers and pesticides; Clitter = cost of chicken litter[2]; Cinorg = cost of conventional fertilizers and pesticides including micronutrients; Ctillage = cost of tillage (bullock and/or tractor); Cwage = cost of labor; and Cpp = postproduction cost of processing, packaging, and transportation.

Gross income is calculated as:

$$\begin{aligned} \text{Gross incomefrom selected crops} = \\ \text{income accrued from own consumption of respective crop} \\ + \text{income accrued from market sell of respective crop} \end{aligned} \qquad (9.2)$$

Net return (Maharjan 1997) is calculated as:

$$\text{Net return} = \text{Gross income} - \text{Cost of production} \qquad (9.3)$$

To compare cost components of production and net return from crop cultivation under these two farming systems, data was analyzed using two-sample t-test. In order to analyze impact of farmers' socioeconomic variables including production factors on crop production, ordinary least square (OLS) model was used.

OLS model can be expressed as:

$$y_i = \beta_0 + x_i \beta_i + \varepsilon_i \qquad (9.4)$$

where i is number of observation, y_i is crop production in kg/ha, x_i is socioeconomic variables, β_0 is coefficient of intercept, β_i is parameter to be estimated, and ε_i is an error term.

Statistical Software – STATA 13 was used to regress OLS for the analysis of production of all three crops together. For simplicity, all redundant variables (with insignificant P-value) through backward elimination method were identified and removed. Thus, it can be said that the empirical specification for this model provided below is the best subset of predictors:

---

[2]Chicken litter in this case is not considered as organic input because of the use of chemically induced feed in poultry farming.

$$\begin{aligned}\text{Production (kg/ha)} = \ & \beta_0 + \beta_1 \text{farm\_method} \\ & + \beta_2 \text{HHHgender} + \beta_3 \text{HHHprimary\_occu} + \beta_4 \\ & \text{LSU} + \beta_5 \text{farm\_size} + \beta_6 \text{membership} + \beta_7 \text{org\_training} \\ & + \beta_8 \text{VDC} + \beta_9 \text{agrovet} + \beta_{10} \text{credit} + \beta_{11} \text{ land} + \beta_{12} \text{seed} \\ & + \beta_{13} \text{ organic\_inputs} + \beta_{14} \text{ chicken\_litter} \\ & + \beta_{15} \text{chemical\_inputs} + \beta_{16} \text{ tillage} + \varepsilon \end{aligned}$$

(9.5)

As per the regression rule, diagnostic tests were carried out. Variation inflation factor (VIF) value of 1.29 (below 10) indicates multicollinearity among the variables does not exist. Breusch-Pagan/Cook-Weisberg has a significant P-value, whereas White's test did not show significant P-value implying that there is problem of linear forms of heteroscedasticity, i.e., variance of error term is not constant. To correct heteroscedasticity of any kind, model estimation was conducted using robust standard errors.

9.5 Production and Net Return of Selected Crops

9.5.1 Production Factors, Production, and Net Return

Among the cost components, only chemical input is such component, which can be calculated only for conventional farming. Besides that all other costs are employed in both farming systems, among which cost of land and seed, wage, and postproduction cost significantly differ between the two. Surprisingly, the average land rent for conventional farming is higher than for organic farming (Table 9.5). During the field survey, respondents did not vary land rent depending on their choice of farming system. Therefore, it is assumed that this variation could be attributed to factors other than the farming system. The study shows that average per hectare cost on seed is significantly higher for conventional than organic carrot and potato. Conventional farmers mostly use hybrid seeds which is much more costly than local varieties, whereas organic farmers mostly save local varieties or buy from their neighbors.

The most common kind of organic input employed for this crop production is livestock manure, which is used by all farmers of both categories, followed by biopesticides and effective microorganism (EM). EM is a combination of useful regenerative microorganisms that improves soil quality. Investment for organic input is significantly higher in organic potato, whereas for chicken litter, it is actually significantly higher for conventional potato. In addition to livestock manure and chicken litter, most of the conventional farmers rely mainly on chemical inputs such as chemical fertilizers (urea, diammonium phosphate (DAP), and muriate of potash (MOP)), followed by micronutrients (vitamin, zinc, and boron),

Table 9.5 Net-return calculation for selected crops under organic and conventional farming systems

Factors (NRs./ha)	Carrot Organic Mean ± SD	Carrot Conventional Mean ± SD	Carrot T-test	Potato Organic Mean ± SD	Potato Conventional Mean ± SD	Potato T-test	Cauliflower Organic Mean ± SD	Cauliflower Conventional Mean ± SD	Cauliflower T-test
Land	12,054 ± 2478	13,012 ± 2402	0.083*	11,696 ± 2664	10,759 ± 1919	0.032**	11,870 ± 2522	10,478 ± 1666	0.012**
Seed	14,639 ± 13,729	20,750 ± 14,879	0.059*	10,813 ± 6421	15,270 ± 4899	0.0001***	13,796 ± 6636	11,690 ± 6605	0.173
Organic inputs	21,387 ± 6362	19,167 ± 6954	0.138	21,533 ± 12,485	16,084 ± 11,031	0.014**	17,209 ± 6642	15,467 ± 8716	0.294
Chicken litter	3406 ± 7287	4114 ± 4114	0.633	332 ± 670	1294 ± 2357	0.002***	3241 ± 6617	2420 ± 5778	0.584
Chemical inputs	–	2243 ± 2569	–	–	13,105 ± 11,547	–	–	30,178 ± 47,114	–
Tillage	17,117 ± 6947	18,497 ± 13,724	0.559	13,184 ± 12,552	14,764 ± 11,418	0.477	8377 ± 10,899	8640 ± 10,181	0.916
Wage	30,028 ± 6411	25,215 ± 7152	0.002***	34,800 ± 24,041	27,818 ± 23,171	0.111	6211 ± 8922	7176 ± 7505	0.630
Postproduction	10,043 ± 3272	12,934 ± 4665	0.002***	7967 ± 6734	10,808 ± 9103	0.053*	229 ± 728	2222 ± 3217	0.000***
Total cost	108,673 ± 22,704	115,932 ± 24,649	0.173	100,319 ± 37,234	109,902 ± 32,739	0.142	60,933 ± 20,532	88,271 ± 54,252	0.0002***
Production (kg/ha)	16,887 ± 4183	19,829 ± 5681	0.009***	9273 ± 3542	12,175 ± 4974	0.0003***	16,121 ± 15,568	19,165 ± 19,208	0.425
Sold (kg/ha)	7995 ± 8886	15,084 ± 10,343	0.001***	35,094 ± 58,767	39,108 ± 61,976	0.718	17,928 ± 55,887	147,272 ± 212,322	0.000***
Sold price (NRs./kg)	11 ± 3	10 ± 2	0.040**	18 ± 2	18 ± 3	0.185	17 ± 5	13 ± 6	0.140
Consumed (kg/ha)	8892 ± 8045	4750 ± 6729	0.016**	131,659 ± 61,540	169,164 ± 79,030	0.004***	224,651 ± 237,984	120,809 ± 161,508	0.047**
Consumed price (NRs./kg)	10	10	–	18	18	–	15	15	–
Gross income (NRs./ha)	178,760 ± 47,414	187,116 ± 58,701	0.481	166,752 ± 63,172	208,272 ± 83,637	0.003***	242,578 ± 233,170	268,080 ± 253,300	0.644
Net return (NRs./ha)	70,086 ± 35,259	71,184 ± 45,658	0.903	66,433 ± 38,677	98,370 ± 68,465	0.002***	181,645 ± 223,663	179,810 ± 214,422	0.972

Source: Field survey (2013)
Note: SD[1] is standard deviation
***1%, **5%, and *at 10% level of significance

9.5 Production and Net Return of Selected Crops

and chemical pesticides (weedicide and insecticide). Investment on chemical input is higher for cauliflower, followed by potato and carrot.

There is no significant difference in tillage cost across all three crops, but still comparatively it is higher in case of conventional farming. This could be because of soil texture under conventional management which gets harder over time due to exposure from various agrochemicals. The higher wage cost in organic carrot confirms with study by Khaledi et al. (2011), which suggests that organic farming is known to demand more labor. Most of the farmers pointed out that weeding takes more labor in carrot production. Since conventional growers relied on weedicide, their labor cost is significantly lower.

Postproduction cost includes cost incurred in the course of processing, packaging, and transporting the product. Total sold amount has a direct impact on cost of packaging and transporting. Since average sold amount for all three crops is higher in conventional compared to organic farming, average postproduction cost for conventional grower is also higher. Thus, the overall cost is higher for conventional farming compared to organic farming on average, and the main factors attributed to it are seed, chemical inputs, and postproduction cost.

Production is also higher for conventional growers compared to organic growers, and it is significant for carrot and potato. Carrot production of 16,887 kg/ha in organic farming and 19,829 kg/ha in conventional farming is higher than average production at district level (14,000 kg/ha), regional level (central Tarai region = 14,000 kg/ha and central region = 13,000 kg/ha), and national level (10,600 kg/ha) as well (MoAD 2013). Potato production of 9273 kg/ha in organic farming and 12,175 kg/ha in conventional farming is lower than average production at district level (18,700 kg/ha), regional level (central Tarai region = 14,313 kg/ha and central region = 14,597 kg/ha), and national level (13,641 kg/ha) as well. Again, cauliflower production of 16,121 kg/ha in organic farming and 19,165 kg/ha in conventional farming is higher than average production at district level (14,000 kg/ha), regional level (central Tarai region = 16,000 kg/ha and central region = 16,000 kg/ha), and national level (15,400 kg/ha) as well (MoAD 2013).

As for the net return, it is again higher for conventional carrot compared to organic carrot but without any significant difference despite having a significantly higher production. This can be attributed to price at which the carrots are sold. Average price for conventional grower is NRs.10/kg, which is significantly lower than for organic growers who sold at average price of NRs.11/kg. However, only 6% of total organic carrots which are produced by organic farmers in a group could be sold through a cooperative in the cities at 9% premium (Table 9.6). Consumed amount is also included in the gross income that is calculated as an average price at which farmers sold (which in this case is at NRs.10/kg). Net return for conventional potato is significantly higher than organic potato, while for conventional cauliflower, it is lower than its organic counterpart but without any statistical significance. Unfortunately, these two crops were not sold at premium price in the cities. Consumed amount for potato and cauliflower is NRs.18/kg and NRs.15/kg, respectively.

Table 9.6 Organic carrots sold through cooperative in Phoolbari VDC (April–May 2012/March–April 2013)

Quantity sold (kg)	Price (NRs./kg)	Total production (kg)*	Sold (%)	Regular price (NRs./kg)	Premium (%)
5000	12	78,407	6	11	9

Source: Field survey (2014)
Note: Total production (kg)* signifies total amount of carrot produced organically by only those (organic) farmers who are members of a cooperative through which they are sold at premium market in other cities

9.5.2 Factors Impacting Crop Production

The *P*-value for the regression model as a whole is highly significant at 1%, which supports existence of relationship of independent variables with dependent variable (Table 9.7). The R^2 value suggests that about 34% of total variation in value of dependent variable is explained by independent variables in this regression equation. Findings show that organic production per hectare is 2535 kg higher compared to conventional crop production, which is significant at 5%. This shows that organic farming does not necessarily result in lower production than conventional farming. Male-headed households have production higher by 2550 kg/ha, significant at 10%, probably because they have more access to resources that improves production. If HHH's primary occupation is farming, there is probability that crop production will decline by 1163 kg/ha compared to those whose primary occupation is in non-farm sector. The only reason for this unexpected relation could be because those whose primary occupation is in non-farm sector are better able to invest income in these crops, thus leading to better production. A unit increase in LSU will lead to 366 kg/ha increase in production. LSU supplies manure which is considered to be important for improving soil fertility and hence the production. A hectare increase in farm size will result in 2300 kg/ha decline in production, significant at 10%. Farm size is an indication of resource-rich farm households, which implies that such households do not feel obliged to improve production, most probably because of no compulsion in producing excess to sell in the market for higher income generation.

Membership increases production by 2084 kg/ha. This means that the formal or informal interaction that takes place among members is inclined toward improving crop production. Contrastingly, one more training related to organic farming has negative relation to production by 300 kg/ha. This implies that training so far conducted do not concern the crops under consideration. For example, during crop-specific training, which is done under FFS, carrot and potato has not been included so far despite of it being one of the most commercially cultivated crops. Farmers in Phoolbari VDC have 1495 kg/ha less crop production than farmers in other two VDCs. A kilometer more distance to agrovet decreases production by 525 kg/ha, significant at 5%. It clearly shows that agrovet provides production enhancing inputs, the farther access to which has negative impact on the production.

9.5 Production and Net Return of Selected Crops

Table 9.7 Result from ordinary least square model for production per hectare of crop production

Variables	Coefficient	P-value
Farm_method	2535.001	0.039**
HHHgender	2550.114	0.092*
HHHprimary_occu	−1163.17	0.309
LSU	365.9175	0.334
Farm_size	−2300.295	0.069*
Membership	2083.6	0.121
Org_training	−299.4609	0.305
vdc	−1495.375	0.306
Agrovet	−525.1708	0.026**
Credit	4270.848	0.196
Land	0.7188612	0.012**
Seed	0.342268	0.000***
Organic_inputs	0.0957952	0.047**
Chicken_litter	0.2208949	0.171
Chemical_inputs	0.2939261	0.000***
Tillage	0.1135464	0.021**
Constant	−6007.566	0.174

Source: Field survey (2013)
Note: ***1%, **5%, and *at 10% level of significance
Number of observations = 307, Prob > F = 0.0000***
$F(16, 290) = 8.50$, $R^2 = 0.3405$, Root MSE = 9127.8

Those who have access to farming credit will produce 4271 kg/ha more crop than those who do not have such credit.

One more rupee investment in land will increase crop production by 0.72 kg/ha, significant at 5%. A rupee increase in seed per hectare increases crop production by 0.34 kg/ha, significant at 1%. Other inputs such as organic inputs, chicken litter, and even chemical inputs increase crop production too. A rupee more investment in organic inputs such as livestock manure, bio-pesticides and/or EM will increase crop production by 0.1 kg/ha, significant at 5%. A rupee more investment in chicken litter will increase crop production by 0.22 kg/ha. A rupee more investment in chemical inputs such as chemical fertilizers and pesticides and micronutrients will increase crop production by 0.29 kg/ha, significant at 1%. A rupee more investment in tillage will increase crop production by 0.11 kg/ha, significant at 5%. This positive impact of production factors on production implies that these inputs have been underutilized, and hence increasing its amount would positively impact the production. Among these production input variables, it is clear that land, seed, and chemical inputs have the highest positive marginal impact on crop production. However, conventional farming is also known to decline soil fertility and create negative repercussion on environment and health of farmers over time (Bhatta and Doppler 2010; Weiss 2004; Shrestha and Neupane 2002), which again makes us question if discarding organic farming in favor of chemical inputs for short-term benefit makes it a worthwhile decision.

9.6 Summary

This chapter analyzed production and net return from three crops, namely, carrot, potato, and cauliflower, which are among the important crops in the study area as well as in the national context. The average production was found to be higher for conventional farming for all three crops, of which carrot and potato have significant difference. Average net return from conventional potato is found to be significantly higher. As for carrot, conventional method and for cauliflower, organic method have better net return, although both are not statistically significant. For all three crops, carrot, potato, and cauliflower combined, organic production was found to be higher than conventional production, as has been shown by the ordinary least square model. This shows that organic farming does not necessarily result in lower crop production. Male-headed households have better crop production because of better access to resources responsible for improving the production. Larger farm size decreases production implying that resource-rich farmers are less obligated to improve production to sell the excess in the market for higher income generation. Farther distance to agrovet will have negative impact on crop production because of difficulty in accessing production-enhancing inputs that are available through agrovets. Investment in land under cultivation of the respective crop, seed, organic inputs such as livestock manure, bio-pesticides, and/or EM; chemical inputs such as chemical fertilizers and pesticides, and micronutrients; and tillage will have positive impact on the production. It means that additional investment in these production factors will result in higher production. Although investment in chemical fertilizers, pesticides, and/or micronutrients results in higher production in the short run, its long-term use is known to deplete the soil fertility which remains outside the scope of this study. Thus, to upgrade the crop production in a more sustainable way, organic inputs should be enhanced.

References

Adil SA, Badar H, Sher T (2004) Factors affecting gross income of small farmers in district Jhang-Pakistan. Pak J Life Soc Sci 2(2):153–155

Ahmad M, Jadoon MA, Ahmad I, Khan H (2007) Impact of trainings imparted to enhance agricultural production in district Mansehra. Sarhad J Agric 23(4):1211–1216

Akal PA (2014) Influence of common interest group approach on orphan crops productivity among farmers, Kenya. Am Int J Soc Sci 3(3):127–132

Badgley C, Moghtader J, Quintero E, Zakem E, Chappell MJ, Avilés-Vázquez K, Samulon A, Perfecto I (2007) Organic agriculture and the global food supply. Renew Agricult Food Syst 22 (2):86–108

Bhatta GD, Doppler W (2010) Socio-economic and environmental aspects of farming practices in the peri-urban hinterlands of Nepal. J Environ Agricult 11:26–39

ESCAP (2012) Organic agriculture gains ground on mitigating climate change and improving food security: healthy food from healthy soil. Economic and Social Commission for Asia and the Pacific (ESCAP)

References

FAO (1996) Technical background document: food and international trade. Economic and social development department. Food and Agriculture Organization of the United Nations, Rome

FAO (2015) Men and women in agriculture: closing the gap. http://www.fao.org/sofa/gender/keyfacts/en/. Retrieved 27 July 2015

Giovannucci D (2005) Organic agriculture and poverty reduction in Asia: China and India focus. International Fund for Agricultural Development

Guo G, Wen Q, Zhu J (2015) The impact of aging agricultural labor population on farmland output: from the perspective of farmer preferences. Math Prob Eng 730618:1–8

Hazell P, Wood S (2008). Drivers of change in global agriculture. Philosophical transactions of the Royal Society B:495–515

IFAD (2013) Smallholders, food security, and the environment. International Fund for Agricultural Development (IFAD), Rome

Kafle B (2010) Determinants of adoption of improved maize varieties in developing countries: a review. Int Res J Appl Basic Sci 1(1):1–7

Khaledi M, Weseen S, Sawyer E, Ferguson S, Gray R (2011) Factors influencing partial and complete adoption of organic farming practices in Saskatchewan, Canada. Can J Agric Econ 58 (1):37–56

Leu A (2011) Scientific studies that validate high yield environmentally sustainable organic systems. Organic Federation of Australia, Mossman

Maharjan KL (1997) Impacts of irrigation and drainage schemes on rural economic activities in Bangladesh. Research Center for Regional Geography. Hiroshima University, Hiroshima

Meisner C (2007) Why organic food can't feed the world? http://www.cosmosmagazine.com/features/online/1601/why-organic-food-cant-feed-world. Retrieved 29 Oct 2012

MoAD (2013) Statistical information on Nepalese agriculture 2012/2013 (2069/070). Ministry of Agricultural Development (MoAD), Kathmandu

MoAD (2015) Ministry of Agriculture Development (MoAD). http://www.doanepal.gov.np/. Retrieved 17 Jan 2015

MoE (2011) Status of climate change in Nepal. Ministry of Environment, Government of Nepal, Kathmandu

Parvin M, Akteruzzaman M (2012) Factors affecting farm and non-farm income of Haor inhabitants of Bangladesh. Progress Agric 23(1–2):143–150

Ramdhani MA, Santosa E (2012) Key success factors for organic farming development. Int J Basic Appl Sci 1(1):7–13

Samriddhi (2011) Commercialization of agriculture in Nepal. Samriddhi – The Prosperity Foundation, Kathmandu

Scialabba NE-H (2007) Organic agriculture and food security. Food and Agriculture Organization of the United Nations, Rome

SECARD (2011) Market oriented organic agriculture promotion project (MOAP) in Chitwan district of Nepal. Society for Environment Conservation and Agricultural Research and Development (SECARD) Nepal, Kathmandu

Shrestha PL, Neupane FP (2002) Socio-economic contexts on pesticide use in Nepal. Landschaftsökologie und Umweltforschung 38:205–223

Tesema SF (2006) Impact of technological change on household production and food security in smallholders agriculture: the case of wheat-tef based farming systems in the central highlands of Ethopia. Cuvillier Verlag, Gottingen

Trewavas A (2002) Malthus foiled again and again. Nature 418:668–670

Vaarst M (2010) Organic farming as a development strategy: who are interested and who are not? J Sustain Develop 3(1):38–50

Weiss J (2004) .Global organics. http://newhope360.com/agriculture/global-organics. Retrieved 16 March 2012

Willer H, Lernoud J (2016) The world of organic agriculture: statistics and emerging trends. International Federation of Organic Agriculture Movements (IFOAM)/Research Institute of Organic Agriculture (FiBL), Bonn/Frick

Chapter 10
Status of Local Organic Market in Nepal

Abstract Organic market is growing globally, and the distance from sellers (usually developing countries) to buyers (developed countries) is getting more evident. This has posed problems of its own such as complications in obtaining and maintaining internationally recognized standards and following the footsteps of conventional model characterized by specialization, capital intensification, export orientation, increased processing, packaging, and long-distance transporting that is controlled by few large corporate retailers. In such scenario, importance of local organic market in developing countries cannot be underrated, which in fact is growing at a steady pace. However, information on its status remains severely lacking, which is why this study is conducted to analyze local organic market in Nepal. The study took place in two major cities of Nepal where presence of organic market is quite evident. A total of 15 categories of products were identified, cereals, vegetables, spices, pulses, oilseeds, fruits, tea, coffee, juice, wine, pickles, honey, jam, snacks, and skin care products, of which vegetables have the highest variety. While most of the organic products are sold on the basis of word of mouth, some are certified by designated agencies, while others are approved by various agencies as safe to consume. The premium is usually higher for certified organic products, but surprisingly some self-claimed organic products are found to have higher premium rate than those certified as well because of already established networks.

10.1 Introduction

Although organic sector is growing in both developing and developed countries, whether it is production or consumption oriented varies significantly among countries or regions depending on the purpose the sector is built on. Currently North America and Europe are the regions with high concentration of market for organic products. Within Asia also market for organic is more confined in affluent countries such as Japan while others have export-oriented sectors (Fig. 10.1) (Willer and Lernoud 2014). Due to concentration of organic market in economically better-off areas of the world, the word "organic" often holds a connotation of rich man's food as it is usually more expensive than conventional food (Belicka 2005; FAO 2014). The prospect of higher income has persuaded developing countries to take part in global organic market. Usually they join in fair trade arrangement through

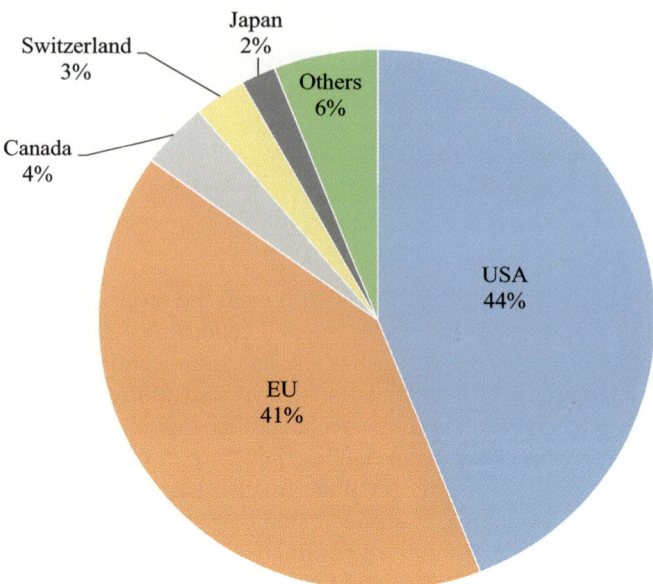

Fig. 10.1 Global market for organic food in 2012 (*Source*: FiBL and IFOAM, 2014)

consolidation with importing countries. Even then smallholder farmers in developing countries can face numerous difficulties in the way of lack of adequate financing, management skills, consistency in workforce, logistics, partnership and cooperation, and cultural differences as a result of globalization (Halberg et al. 2006). On one hand, certification does help farmers to integrate into the global premium market for organic products. But such globally uniform standards, which are usually imposed by developed or importing countries in the north and on developing or exporting countries in the south, might not actually blend with conditions of the south. For example, there is restriction to use neem only in roots of mother plants according to the European Union standard. But in tropical countries where pests can multiply at an alarming rate, it becomes necessary to use them as pesticides in a way that may violate the rule. Obtaining and maintaining internationally recognized standards, high level of record keeping, delay in procuring certification, cost of certification, and annual reinspection are other major obstacles for smallholder farmers. Because of this, often times such standards and control methods rather obstruct the potential growth and spread of the organic sector (Halberg et al. 2006; Barrett et al. 2002; Harris et al. 2001).

Moreover, global organic food market is facing a greater risk of following the footsteps of conventional model characterized by specialization, capital intensification, export orientation, increased processing, packaging, and long-distance transporting that is controlled by few large corporate retailers. Specialization and capital intensification reduce diversity, increase risk of a single crop failure, and limit natural nutrient cycling processes, which could have been achieved through multiple/intercropping method. Market concentration increases vulnerability among farmers in case of price fluctuation or market failure. More so, when it is in the hands of few retailers, there is price monopoly and farmers would no longer have control over price. Growing distance of trade, especially from south to north where developing countries like Brazil, Egypt, and Uganda are now exporting to Europe and North America, has increased ecological footprints as well (Halberg et al. 2006; Kilcher et al. 2008; Knudsen 2010). Thus, commercialization of organic farming has jeopardized the very fundamental elements of organic movement, which is to be self-sufficient and preserve the environmental integrity. Besides the formal organic market with huge discrepancy between countries (primarily developed versus developing), there is still a large share of unaccounted organic area, which prevails mainly in developing countries. Because it takes place outside the formalized market, it is difficult to quantify its extent. Though not certified, it can fetch higher price based on consumers' willingness to pay in a local market, besides providing with other benefits of increase in productivity, saving on purchase of external inputs and transport cost, and getting up close with consumers (FAO 2014; Halberg et al. 2006).

Therefore, local organic market also plays an important role in maintaining environmental integrity and having more control over the products. But information on local organic market, especially in developing countries where local (organic) market is run on informal basis, is very limited to understand the opportunities and constraints therein. In case of Nepal, organic sector is still in its initial stage as has been characterized by many literatures that it lacks proper data, market information, and research-based activities (Bhatta et al. 2009; Pokhrel and Pant 2009). Thus, the overall development of organic market remains quite slow. In such situation, this study was conducted to analyze the general local market scenario for organic products so that present status of organic products available in the local market can be understood to develop it further.

10.2 Data Collection

This study was conducted in December, 2015. About 22 organic outlets situated within Kathmandu metropolitan and Lalitpur sub-metropolitan cities were visited, which are part of Kathmandu and Lalitpur districts, respectively, and are two most densely populated urban centers in Nepal. Out of 26,494,504 denizens of Nepal, Kathmandu metropolitan accounts for 3.68% and Lalitpur sub-metropolitan for 0.83% (CBS 2012).

The main market locations for organic products are situated within these two cities, and it is comprehensible as especially in developing countries like Nepal, organic market is confined within urban areas because of presence of affluent people who have higher education level to understand the benefits of consuming organic products and high income to have ability to pay for its premium price as well. This trend is prevalent in urban areas of other South Asian countries as well such as India (Kilcher et al. 2008; Singh 2013; BIOFACH 2014a, b), Bangladesh (Willer and Lernoud 2014; Sarker and Itohara 2008; Hoque 2012), and Sri Lanka (Willer et al. 2013).

Fig. 10.2 Local organic markets in Nepal (*Source*: Field survey 2015)

Almost all outlets selling organic products, fully or partially, within these two cities were visited (Fig. 10.2). However, since there is no particular reliable source of information to get the list of all such outlets and due to limited knowledge, some

10.2 Data Collection

Table 10.1 List of visited outlets

SN	Name	Products sold	Outlet type	Established	Location
1	Aroma Himalayan Coffee	Organic	Retailer	2015	Lalitpur
2	Ayumi Himalaya Agro Ltd.	Organic	W/R*	2013	Lalitpur
3	Bhatbhateni Supermarket	Organic/conventional	Retailer	–	Kathmandu
4	Bhatbhateni Supermarket	Organic/conventional	Retailer	–	Kathmandu
5	Bhatbhateni Supermarket	Organic/conventional	Retailer	–	Kathmandu
6	Bhatbhateni Supermarket	Organic/conventional	Retailer	–	Lalitpur
7	Farmers Mart	Organic	Retailer	2014	Lalitpur
8	Himalayan Beanz	Organic	Retailer	2009	Kathmandu
9	Himalayan Honey Suppliers Pvt. Ltd.	Organic	W/R	2001	Lalitpur
10	Nepal Tea and Coffee Promotion Center Pvt. Ltd.	Organic	Retailer	2013	Lalitpur
11	Non-Pesticide Vegetables Shop	Organic	Retailer	2010	Lalitpur
12	Nepal Organic Coffee	Organic	W/R	1989	Kathmandu
13	Nepal Tea and Coffee Promotion Center Pvt. Ltd	Organic	W/R	2010	Kathmandu
14	Organic Agro Mart	Organic	Retailer	2013	Kathmandu
15	Organic and Health Food Center	Organic	W/R	2012	Kathmandu
16	Ratomato Organics BBQ Bistro	Organic	Retailer	2012	Kathmandu
17	Saleways	Organic	Retailer	2010	Kathmandu
18	Saleways	Organic	Retailer	–	Lalitpur
19	Satyam Organic Honey Shop	Organic	Retailer	2012	Lalitpur
20	Shop Right Supermarket	Organic	Retailer	2003	Kathmandu
21	The Bee Keeping Shop	Organic	Retailer	1996	Lalitpur
22	Top of the World Coffee	Organic	Retailer	2011	Lalitpur
23	Balkhu Vegetable and Fruit Market	Conventional	Retailer	–	Kathmandu
24	Darsan Salik Fruit Supplier	Conventional	Wholesaler	–	Kathmandu
25	Kalimati Vegetable Market 1	Conventional	Retailer	–	Kathmandu
26	Kalimati Vegetable Market 2	Conventional	Retailer	–	Kathmandu

Source: Field visit (2015)
Note: *W/R stands for Wholesaler/Retailer

outlets may have been missed. To get an idea of how high organic products are priced, four exclusive markets for conventional products and four outlets which sell both organic and conventional products (Table 10.1) were visited. Except for two, most of these organic outlets were established from 2001 onward. Out of them, 50%

are located in Kathmandu metropolitan and another 50% in Lalitpur sub-metropolitan cities.

10.3 Analysis of Local Organic Market in Nepal

From the field survey of local organic market, a total of 15 categories are identified: cereals, vegetables, spices, pulses, oilseeds, fruits, tea, coffee, juice, wine, pickles, honey, jam, snacks, and skin care products (Appendix III). The number of items under each of these categories is provided in Fig. 10.3. Of all the categories, vegetables form the highest number of items (32%), followed by fruits (15%), coffee (12%), honey (8%), and tea (7%). Cereals, spices, and pulses share 6%; juice and pickle share 2%, while oilseeds, wine, jam, snacks, and skin care products all share only 1% of the items. Fig. 10.4 shows the number of items packaged under each category. Almost all products are packaged in a plastic wrapper or container, except for vegetables (only 56% packaged), spices (90%), and fruits (65%).

Figure 10.5 shows organic claim of items under each category. All items under pulses, oilseeds, juice, wine, pickles, jam, snacks, and skin care products are self-claimed. "Self-claimed" is the method of assuring organicness of a product through word of mouth. Organic farming can be a result of either voluntarily excluding inorganic inputs or as a result of organic by default. Organic by default is the term used for farmers practicing organic farming forcibly either because of geographical isolation or due to financial inability to purchase inorganic inputs. Except those, most of the items in other categories are self-claimed too, but certification by an agency to prove its organicness or assurance by a third legal entity is also prevalent. This third legal entity's objective is to approve food item to be safe enough for consumption or associating it with some other feature/s rather than certifying the product as organic. For example, although 95% of cereals are self-claimed, 5% is certified by National Association for Sustainable Agriculture, Australia (NASAA), also known as NASAA Certified Organic (NCO). It is widely acknowledged as a service leader in certification both within Australia and world over. It has accreditation with IFOAM, Japanese Agricultural Standard (JAS), and the United States Department of Agriculture (USDA) organic as well (NASAA 2016). Similarly, 67% of items under vegetables category are identified as self-claimed, 1% is certified by NCO, 25% by Organic Certification Nepal (OCN), and 7% by Vegetable Development Directorate (VDD) of Nepal. OCN is the first private initiative to certify organic agricultural production, wild production, processing, and inputs for production in Nepal (CertAll 2016). Being one of the founder members of Certification Alliance (CertAll – a low-cost one-stop service for locally and internationally recognized organic certification), it offers an internationally accredited inspection and certification services to local clients and operators (Bhat 2009).

10.3 Analysis of Local Organic Market in Nepal

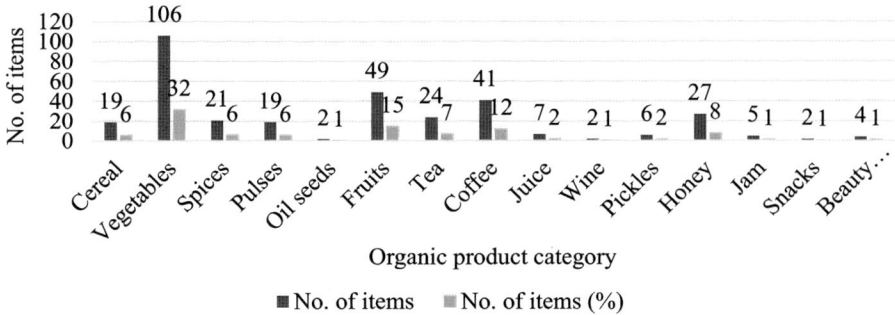

Fig. 10.3 Number of items under each organic product category (*Source*: Field visit, 2015)

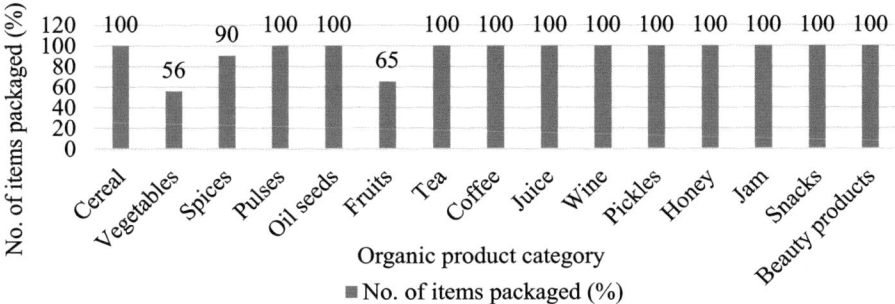

Fig. 10.4 Number of items packaged under each organic product category (*Source*: Field visit, 2015)

VDD is 1 of 12 technical directorates of the Department of Agriculture in Nepal. It serves as a focal point for national and international level institutes for vegetable subsector field. Among its many responsibilities of formulating national policy, strategy, periodic plan, and annual program, preparing guidelines for program implementation, supervising and monitoring district level vegetable program, giving technical support to districts and resource centers, and maintaining necessary information required for and national level data base of vegetable subsector, certifying organic product is not one of them (VDD 2016). This is one of the strategies of sellers to allure buyers by assuring them that it is reinforced by a government institution.

For spices, 10% of it is self-claimed as organic, while 33% is certified under NCO and 19% under OCN. About 67% of fruits are self-claimed as organic, 4% is by NCO, 27% is by OCN, and 2% by VDD. Organic tea and coffee are among the very popular organic products and comparatively have more items certified. In case of tea, only 13% is self-claimed, 4% is ratified by ISO, 25% by NCO, 17% by National Tea and Coffee Development Board (NTCDB), 15% by OCN, and 29% by

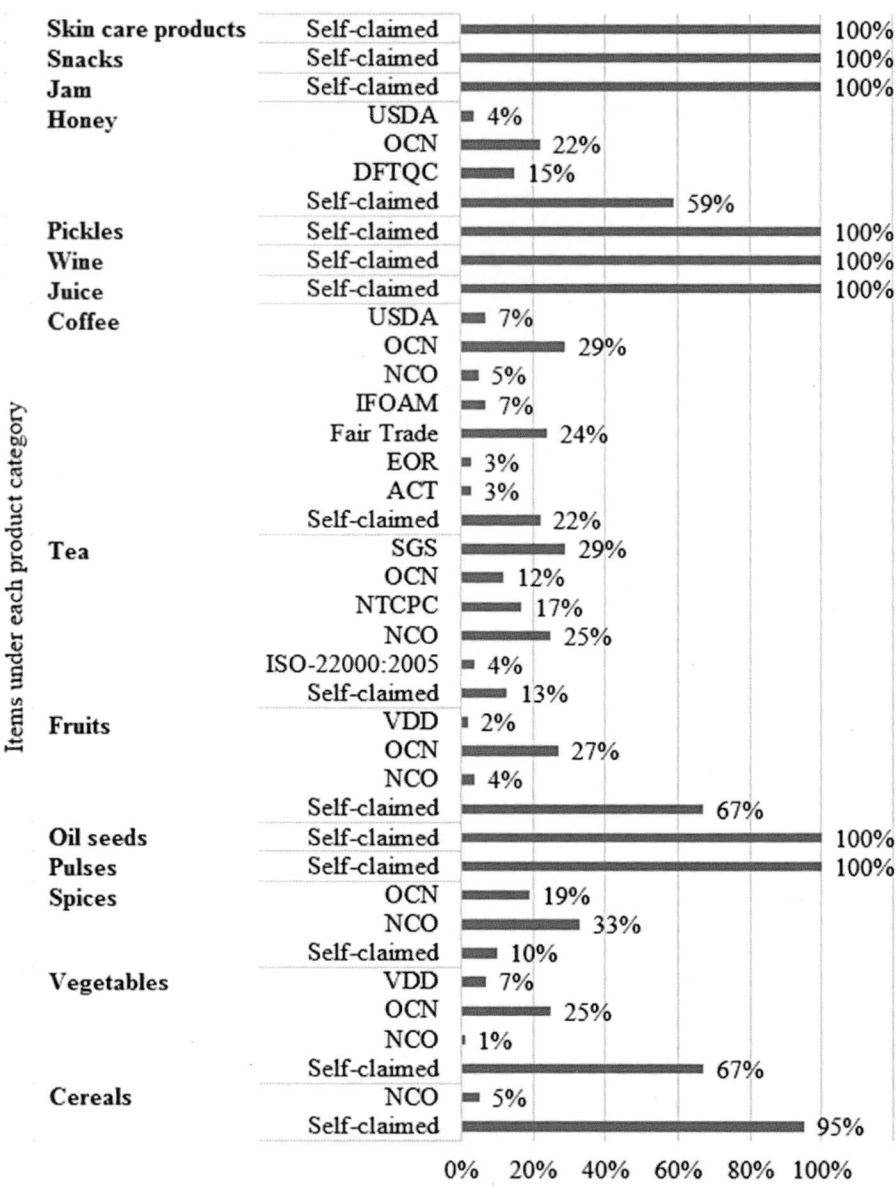

Fig. 10.5 Organic claim of items under each product category (*Source*: Field survey, 2015)

10.3 Analysis of Local Organic Market in Nepal

Société Generale de Surveillance (SGS – an inspection, verification, testing, and certification company in Switzerland). The International Organization for Standardization (ISO) is a worldwide federation of national standard (ISO member) bodies, and ISO 22000:2005 specifies requirements for a food safety management system where all the parties involved in the food chain should exhibit their ability to control food safety hazards for safer food at the time of human consumption (ISO 2016). NTCDB is a commodity board established under the support of Ministry of Agricultural Development (MoAD) to promote and strengthen tea and coffee sector through policy formulation and technical and managerial support (NTCDB 2016).

For coffee, 22% is self-claimed, 3% is certified by Organic Agriculture Certification Thailand (ACT), 3% by European Organic Regulations (EOR), 24% through fair trade, 7% by IFOAM, 5% by NCO, 29% by OCN, and 7% by USDA organic. It should be noted that fair trade exists to ensure better price for producers in developing countries but does not necessarily guarantee that the product is organic (Cierpka 2016). Finally for honey, 59% is self-claimed, 15% is certified under Department of Food Technology and Quality Control (DFTQC), 22% by OCN, and 4% by USDA. DFTQC is established to ensure and enhance quality and safety of food and feed products in Nepal, including appropriate food processing and postharvest techniques to promote agribusinesses (DFTQC 2016).

Figure 10.6 shows source of organic products brought to sell within Kathmandu and Lalitpur cities. There are not many actors involved from the time of production till its point of sell. Usually it is brought directly from farmers to the point of sell. Cereals are brought directly from (group) producers from places such as Godawari (municipality in Lalitpur District), Jumla, Kavre, Mustang, Nawalparasi, and Pokhara districts. Vegetables are brought from Godawari, Thankot (village in Kathmandu District), and districts such as Kavre, Nawalparasi, and Pokhara. Spices, oilseeds, fruits, and other products such as pickles, jams, snacks, and skin care products are brought from Godawari, Kavre, Nawalparasi, and Mustang districts; pulses from Jumla, Kalikot, and Mustang districts; and drinks from Godawari, Ilam, Kavre, Mustang, Nawalparasi, and Palpa districts. Supply chain system in local organic market is quite simple. There are no middlemen but a direct exchange between producer and sellers. Packaging is also undertaken by one of these two actors. In case of smallholder farmers that usually operate in a group, there can be a mediator who is responsible for collecting, packaging, and transporting produces to the sellers. Some producers are also directly involved in production and selling of their produce.

In order to compare the premium rate, certification for this study has been divided into three types: self-claimed, by certifier, and by others (Table 10.2). "By certifier" means certified by recognized certification agencies such as OCN, NCO, SGS, ACT, EOR, IFOAM, and USDA; "by others" means by organizations who did not certify per se, but sellers use their recognition of food products for complying with certain food standard. Such organizations are VDD, ISO, NTCDB, fair trade, and DFTQC. The highest premium rate is 40% for avocado that is certified "by certifier," and the lowest 2% is for beans (white) which is self-claimed as organic. Only lady's finger has same price for both organic and

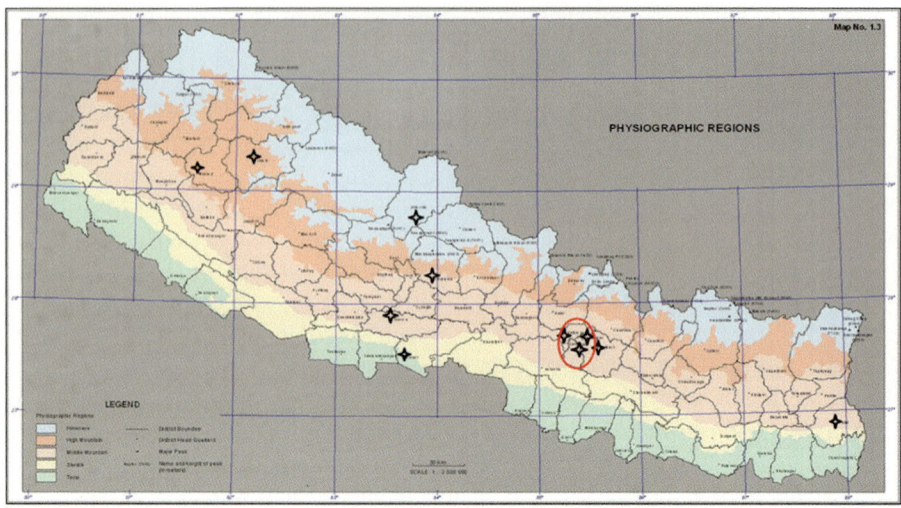

Fig. 10.6 Map of Nepal showing source of organic products and destination of sale (*Source*: http://cbs.gov.np/photo_gallery_photo_gallery_photo_gallery_general. *Note*: The mark in the map indicates source where organic products come from and indicates Kathmandu and Lalitpur districts where organic market exists)

conventional. Only carrot, Colocasia, and onion have higher price for self-claimed organic than certified ones. This can be attributed to the fact that in Nepalese context, some farmers even without certification are able to get premium price purely based on mutual trust or personal links, whereas others are devoid of such benefit despite of being certified because of poor marketing system and skill. Those selling at the premium rate without certification are a result of first mover's advantage where after years of being in the business, these sellers have established their own brand and loyal consumer base. On the other hand, those who are not able to sell their produces even with certification are those who got the financial and technical assistance of certification from an NGO as a way to promote organic farming among smallholder farmers. But unfortunately, these farmers are not given any marketing assistance, and with lack of prior experience and poor networking, they mostly end up selling in the local market with similar price as conventional products. Besides such certification is usually a one-time assistance, which in fact should have been renewed on a yearly basis. But because of lack of fund, these farmers usually do not opt for the certification process on their own. Thus, they are further discouraged to continue their effort to link with the niche organic market (Singh and Maharjan 2013). On an average, premium for cereals is 21%, vegetables is 17%, pulses is 20%, and fruits is 23%.

Organic sellers claim that local demand is definitely increasing, and it is no more confined to foreigners but Nepalese consumers are also equally on rise. Before people used to complain about expensiveness of organic products but now that is changing. Consumers have started to realize the health benefits of organic products

10.3 Analysis of Local Organic Market in Nepal

Table 10.2 Premium price based on organic claim of products

SN	Cereals	(Organic) average price/kg	Certification	(Conventional) average price/kg	Premium %
1	Barley (flour)	130	Self-claimed	100	30
2	Buckwheat (flour)	123	Self-claimed	100	23
3	Millet (flour)	93	Self-claimed	80	16
4	Millet (white)	51	Self-claimed	45	13
5	Rice (basmati white)	150	Self-claimed	130	15
6	Rice (brown)	123	Self-claimed	110	12
7	Rice (brown)	140	By certifier	110	27
8	Rice (red)	120	Self-claimed	100	20
9	Rice (Taichin)	115	Self-claimed	95	21
10	Rice (white)	78	Self-claimed	60	30
11	Wheat (flour)	52	Self-claimed	41	27
SN	Vegetables	(Organic) average price/kg	Certification	(Conventional) average price/kg	Premium %
1	Basil	100	Self-claimed	90	11
2	Beet root	300	Self-claimed	280	7
3	Bitter gourd	105	Self-claimed	80	31
4	Bitter gourd	105	By certifier	80	31
5	Bok choy	50	Self-claimed	48	4
6	Bottle gourd	64	Self-claimed	50	28
7	Bottle gourd	62	By certifier	50	24
8	Brinjal	59	Self-claimed	52	13
9	Brinjal	67	By certifier	52	29
10	Broad bean	47	By certifier	40	18
11	Broccoli	100	Self-claimed	80	25
12	Cabbage	55	Self-claimed	46	20
13	Cabbage	57	By certifier	46	24

(continued)

Table 10.2 (continued)

SN	Vegetables	(Organic) average price/kg	Certification	(Conventional) average price/kg	Premium %
14	Cabbage	60	By others	46	30
15	Carrot	70	Self-claimed	55	27
16	Carrot	68	By others	55	24
17	Carrot	62	By certifier	55	13
18	Capsicum	120	Self-claimed	100	20
19	Capsicum	150	By certifier	120	25
20	Cauliflower	65	Self-claimed	63	3
21	Cauliflower	75	By certifier	63	19
22	Cauliflower	70	By others	63	11
23	Chayote	43	Self-claimed	35	23
24	Chilli (green)	84	By certifier	70	20
25	Chilli (red)	265	Self-claimed	250	6
26	Chilli (red)	275	By certifier	250	10
27	Chilli (dry)	375	By certifier	300	25
28	Chilli (Akhabare)	350	Self-claimed	300	17
29	Colocasia	55	By certifier	50	10
30	Colocasia	65	Self-claimed	50	30
31	Cucumber	80	By certifier	70	14
32	Cucumber	130	Self-claimed	100	30
33	Garden peas	137	Self-claimed	123	11
34	Lady's finger	60	By certifier	50	20
35	Lady's finger	50	Self-claimed	50	0
36	Lettuce	60	Self-claimed	56	7
37	Mushroom (Sitake)	775	Self-claimed	700	11
38	Onion	47	Self-claimed	40	18
39	Onion	43	By certifier	40	8
40	Onion	43	By others	40	8
41	Potato	47	Self-claimed	40	18
42	Potato	51	By certifier	40	28
43	Potato	44	By others	40	10

(continued)

10.3 Analysis of Local Organic Market in Nepal

Table 10.2 (continued)

SN	Vegetables	(Organic) average price/kg	Certification	(Conventional) average price/kg	Premium %
44	Pumpkin	55	Self-claimed	53	4
45	Pumpkin	66	By certifier	53	25
46	Pumpkin	65	By others	53	23
47	Radish	45	Self-claimed	40	13
48	Radish	44	By certifier	40	10
49	Tomato	71	Self-claimed	65	9
50	Tomato	80	By certifier	65	23
51	Tomato	77	By others	65	18
SN	Pulses	(Organic) average price/kg	Certification	(Conventional) average price/kg	Premium %
1	Beans (mixed)	180	Self-claimed	150	20
2	Beans (pink)	205	Self-claimed	150	37
3	Beans (Rice)	120	Self-claimed	90	33
4	Beans (white)	255	Self-claimed	250	2
5	Gram (brown bean)	175	Self-claimed	140	25
6	Horse gram	110	Self-claimed	90	22
7	Kidney bean	197	Self-claimed	150	31
8	Lentil (black)	215	Self-claimed	200	8
9	Lentil (red)	140	Self-claimed	125	12
10	Soybean (black)	170	Self-claimed	150	13
11	Soybean (brown)	105	Self-claimed	90	17
SN	Fruits	(Organic) average price/kg	Certification	(Conventional) average price/kg	Premium %
1	Apple	375	By certifier	291	29
2	Apple	291	Self-claimed	235	24
3	Avocado	280	Self-claimed	250	12
4	Avocado	350	By certifier	250	40
5	Coconut	100	By certifier	75	33

(continued)

Table 10.2 (continued)

SN	Fruits	(Organic) average price/kg	Certification	(Conventional) average price/kg	Premium %
6	Orange	125	Self-claimed	110	14
7	Papaya	100	Self-claimed	85	18
8	Pear	145	Self-claimed	125	16
9	Persimmon	260	By certifier	200	30
10	Persimmon	250	Self-claimed	200	25
11	Pomegranate	340	Self-claimed	290	17
12	Strawberry	320	Self-claimed	300	7
13	Walnut	450	Self-claimed	350	29
14	Watermelon	80	By certifier	65	23

Source: Field survey (2015)
Note: Only some crops are selected for analyzing premium rate

and feel that higher price compensates for this quality. Compared to demand, sellers think that supply is limited, both in quantity and variety. Demand is usually high mostly during festive season and winter, particularly when it comes to tea and coffee. There is a good liking for locally grown organic tea and coffee too than international products. But because of incidences like 2015 Nepal Earthquake and fuel shortage (since June/July, 2015), it has undeniably disturbed the supply chain by making it harder to get enough products. There was also a time when one organic outlet and several other coffee shops were closed down for months because of the earthquake. According to one outlet, owing to fuel crisis, online sales of organic products halted completely because they were incapable of delivering products. As another shopkeeper pointed out, the products used to be supplied for two times/day but now has come down to only once a day. It has also increased cost of production, and as a result, prices of products have increased too. Producers also claim that certification process is too lengthy that demands time, cost, and effort which is discouraging especially for small producers and thus should be made convenient and inexpensive.

10.4 Summary

This chapter analyzed local organic market of Nepal based on field survey. The study is confined within two major cities of Nepal where presence of organic market is quite evident. A total of 15 categories of products were identified, cereals, vegetables, spices, pulses, oilseeds, fruits, tea, coffee, juice, wine, pickles, honey, jam, snacks, and skin care products, of which vegetables have the highest variety. All the products are packaged in some form except for vegetables, spices, and fruits, about less than half of which are sold without the package. All items under pulses, oilseeds, juice, wine, pickles, jam, snacks, and skin care products are self-claimed as organic, which means that they are sold on the basis of word of mouth. For the rest, although most of them are also classified as self-claimed, some are certified by designated agencies such as NASAA Certified Organic (NCO), Organic Certification Nepal (OCN), Société Generale de Surveillance (SGS), and Organic Agriculture Certification Thailand (ACT), while others are approved by various agencies as safe to consume such as Vegetable Development Directorate (VDD), National Tea and Coffee Development Board (NTCDB), International Organization for Standardization ISO-22000:2005, and Department of Food Technology and Quality Control (DFTQC) or are merchandised under "fair trade," but sellers use it nonetheless to market their products as organic. The premium for organic products ranged from 2% to 40%. Surprisingly some self-claimed organic products are found to have higher premium rate than those certified. This can be attributed to the fact that in Nepalese context, some farmers even without certification are able to get premium price purely based on mutual trust or personal links, whereas others are devoid of such benefit despite of being certified because of poor marketing system and skill. On average, premium for cereals is 21%, vegetables is 17%, pulses is 20%, and fruits is 23%. Although sellers opined that popularity of organic products is growing gradually, not just among foreigners but local people as well; natural calamities and political instability has caused damage to its growth. Complication of certification process is what adds to the already challenging situation especially for small organic producers and thus should be simplified, made convenient, and inexpensive.

References

Barrett H, Browne A, Harris P, Cadoret K (2002) Organic certification and the UK market: organic imports from developing countries. Food Policy 27(4):301–318
Belicka I (2005) Organic food: ongoing general aspects. Environmental friendly food production system: requirements for plant breeding and seed production (ENVIRFOOD), Talsi
Bhat BR (2009) Opportunity and challenge of organic certification system in Nepal. J Agricult Environ 10:124–128
Bhatta GD, Doppler W, KC, KB (2009) Potentials of organic agriculture in Nepal. J Agricult Environ 10:1–11
BIOFACH (2014a) India organic: the market place for organic people. BIOFACH-India, Bangalore
BIOFACH (2014b) India: a strong growing organic market. BIOFACH INDIA, Bangalore
CBS (2012) National population and housing census 2011. Central Bureau of Statistics (CBS), National Planning Commission Secretariat, Kathmandu
CertAll (2016) Our partners. http://www.certificationalliance.org/ver1/partners.html. Retrieved 26 Sept 2016

Cierpka T (2016) Organic agriculture and fair trade. https://www.organicconsumers.org/old_articles/ofgu/fair-trade-organic.htm. Retrieved 19 Sept 2016

DFTQC (2016) Services. http://www.dftqc.gov.np/content.php?id=213. Retrieved 29 September

FAO (2014) Organic agriculture: FAQ. http://www.fao.org: http://www.fao.org/organicag/oa-faq/oa-faq5/en/. Retrieved 5 July 2014

Halberg N, Alroe HF, Knudsen MT, Kristensen ES (2006) Global development of organic agriculture: challenges and prospects. CABI Publishing, Oxfordshire

Harris P, Browne A, Barrett H, Cadoret K (2001) Facilitating the inclusion of the resource-poor in organic production and trade: opportunities and constraints posed by certification. Department for International Development (DFID), London

Hoque MN (2012) Eco-friendly and organic farming in Bangladesh: international classification and local practice. Institut für Agrarsoziologie und Beratungswesen der Justus-Liebig-Universität Gießen, Giessen

ISO (2016) ISO 22000:2005 – food safety management systems, requirements for any organization in the food chain. International Organization for Standardization (ISO). http://www.iso.org/iso/catalogue_detail?csnumber=35466. Retrieved 28 Sept 2016

Kilcher L, Eisenring T, Menon M (2008) Organic market development in Africa, Asia and Latin America: case studies and Summarys for national action plans. 16th IFOAM Organic World Congress, Modena

Knudsen MT (2010) Environmental assessment of imported organic products. Department of Agriculture and Ecology, Faculty of Life Sciences, University of Copenhagen, Copenhagen

NASAA (2016) Welcome to NASAA: about us. National Association for Sustainable Agriculture, .Australia (NASAA). http://www.nasaa.com.au/welcome1.html. Retrieved 12 Sept 2016

NTCDB (2016) Home. Nepal Tea and Coffee Development Board. http://www.teacoffee.gov.np/en/index.php. Retrieved 26 Sept 2016

Pokhrel DM, Pant KP (2009) Perspective of organic agriculture and policy concerns in Nepal. J Agricult Environ 10:89–99

Sarker MA, Itohara Y (2008) Organic farming and poverty elimination: a suggested model for Bangladesh. J Organic Syst 3(1):68–79

Singh B (2013) On the brink of an organic revolution. India Brand Equity Foundation (IBEF), Haryana

Singh M, Maharjan KL (2013) Prospect of farmers in generating additional income through organic vegetable farming: a case study in Kathmandu valley and Chitwan district of Nepal. J Inter Develop Cooperat 19(4):37–49

VDD (2016) Introduction. Vegetable Development Directorate. http://www.vdd.gov.np/home-content/1. Retrieved 16 Sept 2016

Willer H, Lernoud J (2014) The world of organic agriculture: statistics and emerging trends. International Federation of Organic Agriculture Movements (IFOAM)/Research Institute of Organic Agriculture (FiBL), Bonn/Frick

Willer H, Lernoud J, Kilcher L (2013) The world of organic agriculture: statistics and emerging trends. International Federation of Organic Agriculture Movements (IFOAM)/Research Institute of Organic Agriculture (FiBL), Bonn/Frick

Chapter 11
Field Experimentation of Vegetable Production

Abstract Organic farming relies on locally available resources and methods. Knowing the best choice among many alternatives would be valuable to increase production under a given environmental condition. Thus, this chapter is based on field experimentation of three crops, viz., carrot, kidney bean, and potato, which were selected for their importance as non-staple commercial crops and significance in their share of daily food consumption, as is known through key informant interview. The study was conducted in Chitwan District of Nepal by associating with a farmers' group for organic farming. About ten plots sized $7.5m^2$ each were allocated for each crop, among which two plots were allocated for experimenting crop production with and without irrigation, two plots for with and without mulching using straws, two plots for with and without self-made bio-pesticides, and the remaining four plots for pruning at the rate 0%, 25%, 50%, and 75%. In all three crops, with irrigation, mulching, and bio-pesticides gave better production than without. Crops without pruning gave better production, and the higher the pruning rate, the less became the production. In all three crops, mulching gave the highest production result, although all plots had higher production than Chitwan District's average, except for pruning which was significantly lower than all other plots in case of all three crops.

11.1 Introduction

Production under organic farming has always been a contested issue. It is known to yield on an average 10–15% less than conventional farming, although these are generally compensated for by lower input costs and higher gross margins (Lotter 2003). Some are of the opinion that feeding through organic farming comes with a huge cost of increasing land area and making available enough organically acceptable fertilizer (Trewavas 2002; Meisner 2007), whereas some have argued from their experiences that it indeed can be a solution to growing food demand and preserving environment, given that proper consideration is taken to fix microbial activities in the soil and intensive natural remedies are followed to boost the production (Leu 2011; Brandt 2007). Irrespective of these claims, general trend over the years have shown that organic farming is gaining popularity world over as a way to produce healthier food (Willer and Lernoud 2016) and should be given importance because of its scope of following sustainable means of food production system (IFOAM 2008).

Organic farming relies on locally available resources such as crop rotation, animal manure, green manure, natural enemies, pest-free plant varieties, companion planting, integrated pest management, etc. to control pests, weeds, and diseases and maintain health of soil and that of all living organisms involved as well. Among the number of available options, it is difficult for farmers to understand which method could be more efficient than the other. Knowing the best choice among many alternatives would prove to be valuable to increase production under a given environmental condition. Thus, this study is done based on field experimentation using four different kinds of production inputs/methods: irrigation, mulching, bio-pesticides, and pruning.

11.2 Field Setup for Experimentation

This study was conducted in coordination with Organic Agriculture Producers' Cooperative of Phoolbari VDC, which is one such formal group established to promote organic farming in Chitwan District. The other two informal groups also exist in the adjoining (Shivanagar and Mangalpur) VDCs. Through such groups, farmers have been conducting Farmer's Field School (FFS). It is one of the regular trainings in which the group usually meet on a weekly basis where they learn-by-doing by assessing one crop at a time from as early as its plantation period till the time of harvest. Farmers usually divide groups to be in charge of growing a certain crop through various organic means such as FYM, bio-pesticides, vermicompost, mulching, and so on. They discuss about the amount of inputs required, problems related to pests and diseases and its management, and finally the amount harvested. Such learning process can take up to 16 weeks for each crop. Through such activity, farmers then try to replicate the most successful method in practice as well. Though this study took place in Phoolbari and Shivanagar VDCs, farmers from Mangalpur VDC were also assimilated for better incorporation of participants (Fig. 11.1). Table 11.1 shows the number of participants in each VDC.

For field experimentation, three crops were chosen: carrot, kidney bean[1], and potato. According to key informant interview, these are among the most significant non-staple crops in the study areas, both in terms of commerciality and their share in daily food consumption. These crops hold significance throughout the country as well. With the production of 26,296 mt in 2488 ha, carrot is the seventh most produced vegetable in Nepal. Similarly, potato is the second highest cash crop

---

[1] In order to maintain consistency with Chap. 9, carrot, cauliflower, and potato were the first choice for field experimentation. However, few years ago cauliflower had already been considered for Farmers' Field School. With the intention that farmers will have the opportunity to learn about new crop, it was replaced by kidney bean.

11.2 Field Setup for Experimentation

Fig. 11.1 Organic disease and pest management, Farmer's Field School (*Source*: Field survey 2015)

Table 11.1 Participants based on gender during field experimentation

VDC	Participants		Total
	Male	Female	
Phoolbari	3	37	40
Shivanagar	2	21	23
Total	5	58	63

Source: Field survey (2015)

produced in Nepal with production of 2,690,421 mt in 64,483 ha (MoAD 2013). While there is no specific information on kidney bean, it is stated to be one of the major pulses grown in central Nepal (Krishna et al. 2012). Also these crops have never been studied through FFS in the study area before. Inputs such as irrigation, mulching, bio-pesticides, and pruning were assessed to see their impact on production of these selected crops. Ten plots of 1.5X5 meters per plot were allocated for each crop. Among these, two plots were assigned for comparative study of crop production with and without irrigation, two plots for with and without mulching, two plots for with and without using bio-pesticides, and finally four plots for pruning at various rates.

Water is the basic necessity for any crop growth. Irrigation is necessary in areas with no or limited rainfall. In this case, because of rain water, a plot was irrigated only for the total of six weeks at the rate of 20 liters/week in an assigned plot.

Mulching helps protect the soil from extreme heat and cold, reduce water loss through evaporation, and prevent weed growth as well (Vinje 2016). Mulching was done using straw of about 15 kg in an assigned plot. Bio-pesticide is meant for controlling pests through nontoxic means and thus has no harmful effect on soil unlike the conventional pesticides (EPA 2013). Bio-pesticides should be prepared using different plants with pesticidal properties such as repellant, anti-feedant, toxicants, and growth inhibitor (SSMP 2014). For this experiment, neem, mugwort (*titepati*), garlic, malbar tree (*ashuro*), tallow tree (*khirro*), chinaberry (*bakaino*), field mint (*pudina*), marigold (*sayapatri/hazari*), chilli (*khursani*), lemongrass, sicklepod (*tapre*), onion, basil (*Tulsi*), ipil-ipil, leaf of sweet flag (*bojho*), stinging nettle (*sisno*), leaves and seeds of Szechwan pepper (*timur*), five-leaved chaste tree (Indian privet/*simali*), leaf of tobacco (*surti*), crofton weed (*banmara*), bank's melastoma (*angeri*), peach (*aaru*), rhododendron (*guras*), pira jhar, leaf of papaya (*mewa*), madar (*aankh*), thorn apple/jimsonweed (*dhaturo*), cow dung, and soap were used. Plant materials were chopped into small pieces and mixed together with other inputs, which were then placed in a plastic drum. Cow urine was mixed at the rate of two liters per kilogram of solid material. Three drums of 50 liters capacity were used. The drum was closed as airtight as possible, put in a shady place, and stirred with a wooden stick for about 2–3 times within 20–22 days (since it was during winter, it took more time compared to just about 15 days in summer), after which it was ready to be used. While applying, the prepared bio-pesticide was then diluted with water at the ratio of 1:10 (one part pesticide solution: ten parts water). It was applied every week with the sprayer till the harvest time and the ratio was progressively adjusted to 1:8, 1:6, and 1:4 as the crops matured.

Pruning is an act of cutting off crop's growth that is not useful so that the plant's energy can be directed to growing the best fruit. It is known to control plant size, set direction of plant growth, improve light and air flow around the plant, lessen the threat of pests and disease, and improve crop quality (Hull 2016). But study by Kanyomeka and Shivute (2005) mentions a conflicting result from two studies where one showed increased yield and quality and the other reduced yield and/or quality or no effect at all on tomato production. In the study itself, tomato pruning caused marginal reduction in tomatoes yield, but unpruned tomatoes seem to be more prone to pest attack in which case it was advisable to prune tomatoes in order to minimize the pest attack. In order to compare the result on crop production from no pruning with various rates of pruning, this experiment laid out four plots which were used for pruning at the rate of 0%, 25%, 50%, and 75% to see its impact on overall crop production. The first plot was left untouched (0%); the crop in the second plot was divided into four parts out of which only one part was pruned (25%). Likewise, in the third plot, only half of the crop was pruned (50%), and finally on the fourth plot, three-quarters of crop were pruned. Since the time the crop germinated and throughout its development, the top of weaker/smaller shoots and dead/infected leaves were pruned.

Table 11.2 provides information on plantation of the selected crops. After the plantation, each of the participants gathered 11 times thereafter in a gap of every seven days to study the growing phase of the said crop before it was harvested.

11.2 Field Setup for Experimentation

Table 11.2 Crop plantation information

Plot no.	Treatment plots	Crops					
		Carrot		Kidney bean		Potato	
		Subject	Area	Subject	Area	Subject	Area
1	I	With irrigation	1.5X5 meters	With irrigation	1.5X5 meters	With irrigation	1.5X5 meters
2		Without irrigation	1.5X5 meters	Without irrigation	1.5X5 meters	Without irrigation	1.5X5 meters
3	II	With mulch	1.5X5 meters	With mulch	1.5X5 meters	With mulch	1.5X5 meters
4		Without mulch	1.5X5 meters	Without mulch	1.5X5 meters	Without mulch	1.5X5 meters
5	III	With bio-pesticides	1.5X5 meters	With bio-pesticides	1.5X5 meters	With bio-pesticides	1.5X5 meters
6		Without bio-pesticides	1.5X5 meters	Without bio-pesticides	1.5X5 meters	Without bio-pesticides	1.5X5 meters
7	IV	Pruning at 0%	1.5X5 meters	Pruning at 0%	1.5X5 meters	Pruning at 0%	1.5X5 meters
8		Pruning at 25%	1.5X5 meters	Pruning at 25%	1.5X5 meters	Pruning at 25%	1.5X5 meters
9		Pruning at 50%	1.5X5 meters	Pruning at 50%	1.5X5 meters	Pruning at 50%	1.5X5 meters
10		Pruning at 75%	1.5X5 meters	Pruning at 75%	1.5X5 meters	Pruning at 75%	1.5X5 meters
Seed type		New Koroda		Hetaude		Seto Golo	
Seed amount/plot		100 gm/plot		150 gm/plot		4 kg/plot	
Sowing date		December 2, 2015		December 2, 2015		December 2, 2015	
Harvest date		March 14, 2016		March 27, 2016		March 10, 2016	

Source: Field survey (2015)

Table 11.3 Weekly temperature during crop growing phase

Date	Temperature
Thursday, December 31, 2015	18 °C
Thursday, January 7, 2016	19 °C
Thursday, January 14, 2016	19 °C
Thursday, January 21, 2016	18 °C
Thursday, January 28, 2016	15 °C
Thursday, February 4, 2016	19 °C
Thursday, February 11, 2016	21 °C
Thursday, February 18, 2016	25 °C
Thursday, February 25, 2016	28 °C
Thursday, March 3, 2016	25 °C
Thursday, March 10, 2016	29 °C
Average temperature	22 °C

Source: Field survey (2015)

Temperature on a particular day of study is provided in Table 11.3. Average temperature throughout the growing phase of the crop was 22 °C.

11.3 Result from Field Experimentation

11.3.1 Carrot

Table 11.4 shows result from field experimentation of carrot production. In plot I, production of carrot was higher with irrigation than without irrigation. With irrigation, production was 38,000 kg/ha compared to 36,007 kg/ha without irrigation. Mulching also resulted in higher carrot production than without mulching in plot II. With mulching, production was 42,420 kg/ha compared to 35,733 kg/ha without mulching. Similarly in plot III, application of bio-pesticides too gave higher carrot production than without using bio-pesticides. With bio-pesticides, production was 41,427 kg/ha compared to 34,940 kg/ha without bio-pesticides. Finally in plot IV, pruning at 0% gave the best result than pruning at higher rates. At 75% pruning, production was lowest at 6600 kg/ha compared to 6933 kg/ha at 50%, 9633 kg/ha at 25% and 9733 kg/ha at 0% pruning. Among all the plots, the highest production was the one with mulch, but all plots gave production higher than the average production of carrot in Chitwan District, i.e., 14,000 kg/ha (MoAD 2013), except for pruning for which it was considerably lower (Fig. 11.2).

11.3 Result from Field Experimentation

Table 11.4 Result from carrot production

Treatment plots	Activities	Carrot with leaves (kg/7.5 m^2)	Only carrot (kg/7.5 m^2)	Only carrot (kg/ha)
I	With irrigation	53.7	28.5	38,000
	Without irrigation	48.05	27.01	36,007
II	With mulch	62.29	31.82	42,420
	Without mulch	52.11	26.8	35,733
III	With bio-pesticides	49.4	31.07	41,427
	Without bio-pesticides	53.85	26.21	34,940
IV	Pruning at 0%	14.9	7.3	9733
	Pruning at 25%	14.5	7.23	9633
	Pruning at 50%	12.05	5.2	6933
	Pruning at 75%	10.55	4.95	6600

Source: Field survey (2015)

Fig. 11.2 Women farmers inspecting plot for carrot production (*Source*: Field survey 2015)

11.3.2 Kidney Bean

Table 11.5 shows result from field experimentation of kidney bean production. In case of kidney bean, out of total amount harvested, some were green in color, some

Table 11.5 Result from kidney bean production

Treatment plots	Activities	Total beans (kg/7.5 m²)	Green beans (kg/7.5 m²)	Small beans (kg/7.5 m²)	Unsuccessful beans (kg/7.5 m²)			Good beans (kg/7.5 m²)	Good beans (kg/ha)	Total beans (kg/ha)
					Total	Lost due to pest	Lost due to disease			
I	With irrigation	2.1	0.076	1.36	0.64	0.351	0.293	0.016	21	2800
	Without irrigation	1.9	0.071	1.22	0.60	0.299	0.299	0.015	20	2533
II	With mulch	2.6	0.059	1.77	0.75	0.334	0.418	0.019	25	3467
	Without mulch	2.15	0.077	1.34	0.72	0.329	0.395	0.013	17	2867
III	With bio-pesticides	1.98	0.065	1.55	0.34	0.153	0.191	0.016	21	2633
	Without bio-pesticides	1.45	0.079	0.92	0.44	0.180	0.256	0.012	16	1933
IV	Pruning at 0%	0.48	0.025	0.1	0.03	0.011	0.014	0.33	440	640
	Pruning at 25%	0.47	0.004	0.1	0.02	0.012	0.011	0.34	453	627
	Pruning at 50%	0.44	0.011	0.09	0.04	0.026	0.017	0.3	400	587
	Pruning at 75%	0.31	0.014	0.04	0.02	0.009	0.011	0.24	320	413

Source: Field survey (2015)

11.3 Result from Field Experimentation

Fig. 11.3 Farmers gathered for studying plots for kidney bean. *Source*: Field survey (2015)

were smaller in size, and some were lost to pests and disease. Overall the production of total beans in plot I was 2800 kg/ha for irrigated plot which is higher than 2533 kg/ha for unirrigated plot. In plot II, one with mulch also gave higher production at 3467 kg/ha compared to plot without mulch at 2867 kg/ha. Production on plot III was higher for the one using bio-pesticides at 2633 kg/ha compared to 1933 kg/ha in plot without bio-pesticides. Lastly in plot IV, production was highest at 0% pruning (640 kg/ha), followed by 627 kg/ha at 25%, 587 kg/ha at 50%, and only 41 kg/ha at 75% pruning. Like carrot, production was higher with irrigation, mulching, and bio-pesticides compared to without and even pruning at 0% gave the best result than pruning at higher rates. Again, the overall highest production was the one with mulch, and all plots gave production higher than the average production of kidney bean in Chitwan District, i.e., 1755 kg/ha (Bhandari et al. 2015) except for pruning, for which it was substantially lower (Fig. 11.3).

11.3.3 Potato

Table 11.6 shows result from field experimentation of potato production. In plot I, potato production for irrigated plot was higher at 15,000 kg/ha compared to unirrigated plot at 11,600 kg/ha. In plot II, mulching gave higher production of

Table 11.6 Result from potato production

Treatment plots	Activities	Lost to pests (kg/7.5 m^2)	Total production (kg/7.5 m^2)	Total production (kg/ha)
I	With irrigation	0.45	11.25	15,000
	Without irrigation	0.5	8.7	11,600
II	With mulch	0.7	12.7	16,933
	Without mulch	0.55	9.7	12,933
III	With bio-pesticides	0.3	9.2	12,267
	Without bio-pesticides	0.6	8.7	11,600
IV	Pruning at 0%	0.15	1.85	2467
	Pruning at 25%	0.05	1.85	2467
	Pruning at 50%	0.15	1.25	1667
	Pruning at 75%	0.1	1.12	1493

Source: Field survey (2015)

potato at 16933 kg/ha compared to 12,933 kg/ha for plot without mulching. In plot III, one with bio-pesticides had higher production at 12,267 kg/ha than without at 11,600 kg/ha. Finally in plot IV, potato production was same with no pruning and at 25% which gave the highest production of 2467 kg/ha, followed by pruning at 50% with 1667 kg/ha and at 75% with 1493 kg/ha. Like carrot and kidney bean, production was higher with irrigation, mulching, and bio-pesticides compared to without, and pruning at 0% and 25% gave the best result than at other rates. Similarly, the overall highest production was the one with mulch, and all plots gave production higher than the average production of potato in Chitwan District, i.e., 18,700 kg/ha (MoAD 2013) except for pruning, for which it was again considerably lower (Fig. 11.4).

11.4 Summary

The study did field experimentation of three crops, viz., carrot, kidney bean, and potato, which were selected for their importance as a non-staple commercial crops and significance in their share of daily food consumption. These crops hold significant share in national crop production as well. About ten plots sized 7.5m^2 each were allocated for each crop to do experimentation of respective crop production. Among these, two plots were allocated for experimenting crop production with and without irrigation, two plots for with and without mulching using straws, two plots for with and without self-made bio-pesticides, and remaining four plots for pruning at the rate 0%, 25%, 50%, and 75%. In all three crops, irrigation, mulching, and bio-pesticides gave better production than without. Crops without pruning gave better production, and the higher the pruning rate, the less became the

Fig. 11.4 Farmers weighing potatoes after harvesting from the plots (*Source*: Field survey 2015)

production. In all three crops, mulching gave the highest production result, although all crops had higher production than Chitwan District's average production rate, except for pruning which was significantly lower than all other plots in case of all three crops.

References

Bhandari NB, Kunwar M, Parajuli K (2015) Average production cost and profit margin of pulse, oilseed, spice and commercial crops 2071/2071 (2014/2015). Ministry of Agriculture Development (MoAD), Lalitpur

Brandt K (2007) Organic agriculture and food utilization. Newcastle University, Tyne and Wear

EPA (2013) Biopesticides. United States Environmental Protection Agency (US-EPA), Washington, DC

Hull C (2016) .Pruning vegetable plants: Hands-on your plants, literally. http://www.organicauthority.com/pruning-vegetable-plants-your-hands-on-how-to-guide/. Retrieved 15 June 2016

IFOAM (2008) Definition of organic agriculture. IFOAM – organics international: cultivating change. http://www.ifoam.bio/en/organic-landmarks/definition-organic-agriculture. Retrieved 20 March 2014

Kanyomeka L, Shivute B (2005) Influence of pruning on tomato production under controlled environments. Agric Trop Subtrop 38(2):79–83

Krishna VV, Mehrotra MB, Teufel N, Bishnoi DK (2012) Characterizing the cereal systems and identifying the potential of conservation agriculture in South Asia. The International Maize and Wheat Improvement Center (CIMMYT), Mexico

Leu A (2011) Scientific studies that validate high yield environmentally sustainable organic systems. Organic Federation of Australia, Mossman

Lotter DW (2003) Organic agriculture. J Sustain Agric 21(4):1–63

Meisner C (2007) Why organic food can't feed the world?. http://www.cosmosmagazine.com/features/online/1601/why-organic-food-cant-feed-world. Retrieved 29 Oct 2012

MoAD (2013) Statistical information on Nepalese agriculture 2012/2013 (2069/070). Ministry of Agricultural Development (MoAD), Kathmandu

SSMP (2014) A handbook of technologies and extension approaches from the sustainable soil management programme. Sustainable Soil Management Programme (SSMP), HELVETAS Swiss Intercooperation Nepal, Kathmandu

Trewavas A (2002) Malthus foiled again and again. Nature 418:668–670

Vinje E (2016) .Planet Natural. www.planetnatural.com/garden-soil/. Retrieved 29 Aug 2016

Willer H, Lernoud J (2016) The world of organic agriculture: statistics and emerging trends. International Federation of Organic Agriculture Movements (IFOAM)/Research Institute of Organic Agriculture (FiBL), Bonn/Frick

Chapter 12
Organic Farming from Perspective of Three Pillars of Sustainability

Since the beginning of the twenty-first century, the term "sustainability" has been gaining worldwide attention, not just in regard of natural environment but in far more complex entities such as human societies, cultural traditions, or social institutions. The inclusion of this concept in case of farming sector became prevalent when it was realized that green revolution might not be the ideal way to solve the issues of food insecurity, resource degradation, and regional imbalance of benefits. There is no doubt that the green revolution, which commonly is also known as conventional farming, profoundly increased crop production with the use of chemical fertilizers and pesticides and high-yielding crop varieties and contributed significantly in reducing world hunger. But overtime we have started to realize that such production system has its own share of drawbacks in the form of environmental degradation, health implication, and imbalance of benefit entitlement across regions. Excessive and inappropriate use of chemical fertilizers and pesticides polluted groundwater, streams, rivers, and oceans; triggered land degradation through soil erosion; and severed deterioration of the arable soil. It caused professional hazard; killed beneficial insects and other wildlife; reduced biodiversity; increased pest adaptation and resistance, desertification, and water eutrophication; and affected those who consumed it through food residue.

Sustainability of a farming system means maintaining the quantity and quality of agricultural produce over a considerably longer period of time without an exhaustion of ecosystem services that crops and animals depend on for their productivity. At present, sustainability is widely viewed from the perspective of social, economic, and environmental dimensions. Organic farming is known to be one of the most sustainable forms of production methods. It excludes the use of chemical fertilizers and pesticides, genetically modified organisms, growth regulators, livestock feed additives, or hormones. Instead it relies on natural processes such as crop rotation, animal manure, green manure, natural enemies, pest-free plant varieties, companion planting, integrated pest management, and so on to control pests, weeds, and diseases and maintain health of soil and that of all living organisms involved as well. It emphasizes on the use of locally available resources and optimum production under given environmental condition. It is also known to adapt, mitigate, and being resilient to changing climate, which is a burgeoning issue in present-day scenario. While there have been studies showing benefits of organic farming with regard to all three dimensions of sustainability, its share remains a meager 1% only. On the contrary, it is attributed for producing less compared to its conventional counterpart, is labor intensive, and often times difficult to get hold of enough organically acceptable inputs as prescribed in guidelines of the said country. Another reason is organic food is generally more expensive which makes it less desirable by those with low income. However, any judgment toward feasibility of organic farming in a given society and ways to overcome barriers for expansion of this sector should be context specific, as there have been evidences of benefit provided by organic farming on social, economic, as well as environmental grounds.

While organic farming is surrounded by number of issues, its growth world over has been consistently increasing over the years. By 2014, 172 countries were formally involved in this sector through complying with certain standard and managed by more than two million producers. Its global sales reached US$80 billion in 2014, an increase of nearly 188% compared to more than a decade earlier. Even in South Asia, a region mostly known for being economically backward, has an organic sector that is basically export oriented. However, its local market also seems to be on rise given increasing purchasing power and awareness of health impact of food residues among consumers.

With the understanding that the success of organic farming is very much location specific, characterized by the local social, economic, environmental, and geopolitical scenario, this book tried to do an in-depth analysis of sustainability of organic farming in Nepal. Nepal is a country with a very diverse agriculture sector due to the range of agro-climate brought on by its vast variation in topography and altitude. Though agriculture is one of the major economic sectors of Nepal and despite of it directly engaging more labor force than any other sector, it is highly underdeveloped. Besides being highly weather dependent, other drawbacks of this sector are low investment leading to sluggish development and transfer of technology, subsistence and scattered farming, poor linkage with research and extension, limited subsidies and support in agricultural inputs, rapid depletion of natural resources (declining soil fertility, loss of land due to landsliding, erosion, and deforestation), labor scarcity, and limited trained human resources. The mountain and hill regions have limited agricultural land for crop production, while the plain area on the southern side, also known as Tarai, has a subtropical climate and covers 15% area of the country. It has high potentiality of commercial farming production and is known as "the food bowl" of the country.

Commercialization of farming through conventional means is much prioritized for overall economic development of Nepal, as has been reflected in various agricultural policies. But declining soil fertility, negative repercussions on environment and health of farmers due to use of agrochemicals, and market demand reinforced the organic movement in Nepal. As of 2014, the total share of organic to overall agricultural land including in-conversion areas is reported to be only 0.2%

(9361 ha). However, this is a highly underreported data as much of organic farming has not been formalized yet. In high mountain and middle mountain areas, farmers still rely on traditional knowledge and locally available resources and are largely claimed as "organic by default." Organic farming is claimed to have huge potential in Nepalese context due to exclusion of costly agrochemicals, ecological diversities, and higher labor availability in farming sector, thus making it self-reliant.

The government has also realized the benefits of organic farming, as has been mirrored in growing number of organic farming-friendly policies in recent years. Among various policies made by the government in its favor, the most significant achievement that laid the foundation for organic farming development is National Standards of Organic Agriculture Production and Processing 2007. It is a government-competent organic labeling regulation that provides rules, regulations, guidelines, and procedure for the production and processing of organic products and regulates, monitors, and evaluates organic farming development in Nepal. However, policy implementation is less than satisfactory. Lack of adequate and integrated research, extension, manpower and other support on organic farming production, marketing, and input supply have hindered the development of organic farming promotion. Organic product legislation, standardization, certification, and infrastructure development are also major concerns. Basically the policy provisions are too broad without clear pathways to translate them into actions.

Generally in rural areas of developing countries, there is often a vague difference between organic and conventional farming systems in a sense that both incorporates integrated farming of crops and livestock, unlike in developed countries where conventional farmers are known to have mono-cropping farming system and mostly relies on chemical inputs. Thus, in order to demarcate the actual difference between organic and conventional farming systems, this book takes the case study of Chitwan district where indiscriminate use of agrochemicals is quiet common, but group conversion to organic farming also exists in three adjoining village development committees (VDCs): Phoolbari, Shivanagar, and Mangalpur. Thus, these three VDCs were considered to be an ideal case for this book. Respondents were selected by stratifying individual households based on their membership in a group formed for the purpose of organic farming. From the survey, it was realized that not all farmers belonging to such group are practicing organic farming. Likewise, not all farmers not belonging to such group are practicing conventional farming. Therefore, there are both kinds of farmers within and outside such group, although most organic farmers are group members. Final data of 285 households was used for the analysis.

In the study area, large-scale adoption of organic farming is made possible through the group formation through which farmers are provided training and other extension services. However, rigorous practice of organic farming put forth by the national standard is not practiced extensively as it would take enormous resources. Thus, this study simply segregates organic and conventional farming as the one that restricts and uses chemical fertilizers and pesticides, respectively. However, closer analysis suggests that overtime some farmers go back to practicing conventional farming and some farmers not belonging to such group are also practicing organic farming which for them is following a traditional farming that their predecessors used to practice. Conventional farming in this particular book does not imply that farmers rely on chemical fertilizers and pesticides extensively but occasionally in the event of pests, disease, or weed outbreak, especially in case of vegetables and widely in case of cereal crops which are grown in a larger area. It was found that sometimes crop variety is also to blame for not being able to grow organically. For example, some farmers could grow certain variety of rice and potato without the use of pesticides, but unfortunately many farmers were unaware of it or could not have access to it. Thus, from this particular incident, it can be suggested that support for full conversion can partly be done by distributing organically feasible varieties.

As far as socioeconomic situation of the study area goes, it is mainly patriarchal-based as is predominant in case of Nepalese society. Most of the households have up to secondary level education, and many households consider farming as their primary education. While renting and mortgaging farm land is prevalent, most of the farm is owned by the farmers themselves, and majority of them own livestock. Livestock holding, membership in a group formed for the purpose of organic farming, and training related to organic farming are significantly higher among organic farmers. Taking credit for farming purpose is not so common among farmers, while selling crops in the market is done by majority of the households, but only few are aware of the actual price consumers pay for it. Conventional farmers earn higher farm cash income than organic farmers although there is no significant differences for household heads' age, education, labor availability, livestock holding, farm size, overall farm valuation, non-farm income, distance to significant establishments like agrovet and market, commercialization rate, and crop diversity.

Analysis of group formed for the purpose of organic farming showed that some farmers got more health conscious after being a member of such group as a result of which some also tend to grow same crop organically for home consumption and conventionally to sell in the market. While selling of organic produce through such group in the cities at premium price is very limited, occasional financial and material assistances encourage some farmers to stay within the group by attempting to farm at least minimum land area required by such group to be practiced as organic. Thus, forming such group could be an efficient tool to introduce organic farming on a larger scale. However, it comes with various other challenges of unequal member participation in group activities, entitlement by each member to the benefit received in the form of those assistances, and accessibility to premium market. Smaller interest rate from savings compared to other financial institutions has also diminished the appeal of remaining a member in such groups. While many challenges are difficult to overcome, one of the ways to improve such situation could be to link with retailers whose specialized shops have begun to thrive very recently in the local area, which can hopefully accommodate selling organic products from more farmers.

To study sustainability of a farming system, this book first analyzed households' socioeconomic factors impacting adoption of a farming system to identify which factor encourages or deters adoption of organic farming for policy implication. In any adoption studies of agricultural innovations, socioeconomic factors are as important as agroecological variables and farmers' perception. The result from binary logistic model shows that older farmers are less likely to adopt organic farming. Thus, while introducing organic farming, older farmers should not be prioritized as their capacity to supply labor diminishes which is incompatible for this labor-intensive farming system. Additionally, benefit from organic farming materializes only after few years of conversion, thus diminishing their enthusiasm, as they will be retired soon in the near future which leaves them with less time to enjoy the benefit. Establishment of a group for the purpose of organic farming creates a foundation for commencing organic farming in a group, but it is training that complements the technical knowledge required to continue practicing organic farming in a long run. Organic farming is not just following the traditional way of farming but assimilating them with modern scientific knowledge as well. Rather than relying completely on nature, organic farmers use all the knowledge, techniques, and materials available to work with nature. Thus, training ultimately plays crucial role in knowledge generation and information dissemination and hence causes higher organic farming adoption rate among farmers over a longer period of time. Compared to other two VDCs, farmers in Phoolbari VDC have higher probability of taking up organic farming because of group formation over a longer period of time, higher number of trainings conducted, and other unobserved characteristics. This study shows that farmers in other VDCs need more support and attention in their effort to practice organic farming. Agrovets also sell packaged organic fertilizers and bio-pesticides which could be the reason why farther distance to it results in lesser chance of practicing organic farming, indicating the importance of commercially available organic inputs for the vitality of organic farming.

In order to evaluate environmental implication of these two farming systems, their soil properties have been assessed. Soil has a dominant role to play in improving crop productivity by infiltrating water and air, allowing plant roots to expand and encouraging biota to thrive that in turn controls physical degradation of soil and cycles nutrients based on plant needs. Organic means is known to enhance or sustain overall quality and health of soil ecosystem, while conventional means such as chemical fertilizer depletes soil fertility by reducing soil structure and soil aggregation, decreasing water infiltration, and increasing soil bulk density, soil salinity, nitrogen leaching, and groundwater contamination. Comparison is made based on soil texture, pH, organic matter, and three main mineral components – nitrogen, phosphorus, and potassium – that make up for the healthy soil. About 30 samples, 15 belonging to each farming system, were tested in Department of Soil Test and Service – the national soil testing facility in Nepal. All the variables were analyzed using t-test. Soil texture triangle was used to know the soil type based on proportion of soil particles: sand, silt, and clay; and rest of the variables were measured based on the desired level suggested by Nepal Agricultural Research Council. It was found that there is no significant difference in the soil texture, pH value, organic matter, and nitrogen level. Overall soil texture is sandy loam, which means it has good water infiltration, aeration, and workability but poor nutrient and water-holding capacity. Thus, plants grown in this type of soil will require regular fertilization and irrigation. Overall average value of soil pH is 5.39, which suggest soils tested are moderately acidic. Since pH value closer to seven is desirable for healthy plant growth, amendments such as lime, manure, and moderate amount of wood ash are recommended to raise soil pH. With an average value of 2.68%, according to Nepal Agricultural Research Council's requirement, organic matter in the tested sample is available at the medium level. Compost, manure, or mulching by using straw, hay, grass clippings, and shredded bark are suggested to raise organic matter. Overall average nitrogen level of 0.12% shows that it is also available at the medium level. Although phosphorus and potassium level are comparatively higher in organic soil, overall they are very high in both kinds of soils. In order to avoid or lessen excess amount of phosphorus and potassium, certain measures can be taken such as growing forage crops, legumes, and certain vegetables; making compost from on-farm organic materials; using nitrogen-only fertilizers or with a high nitrogen analysis relative to phosphorus and potassium; and nutrient budgeting. However, there are added responsibilities or challenges associated with these practices. While in the present scenario there seem to be not much difference in organic and conventional soil, same cannot be implied for the future as environmental change takes place after many years of accumulated processes.

Another component included under environmental implication of these two farming systems is adoption of organic means of crop management practices. It has further been divided into two categories: soil fertility and pest management practices. Soil fertility management is usually related to dynamic properties of soil which can be controlled by humans. This study classifies five types of soil fertility management practices: mulching (conserves moisture, protects plant roots, reduces weed growth, improves soil health and fertility), compost-shed (preserves compost/manure pile from volatilization by the sun or leaching by rainfall and maintains nutrient availability), bio-slurry (revitalizes soil fertility), plastic cover (maintains soil moisture and subsequently makes nutrients available), and vermicompost (improves soil fertility). For organic means of pest management, bio-pesticide is taken as it manages pests without having to rely on harmful chemical pesticides that degrades soil over time and increases pests' resistance. This study shows that even though there is influx of modern inputs like chemical fertilizers, pesticides, and micronutrients, conventional farmers still incorporate all organic means of soil fertility and pest management practices analyzed in this study. Although adoption rate for all of such practices is higher among organic farmers, indicating that organic farmers are keener on adopting such practices, especially when it comes to mulching and constructing compost-shed.

Analysis from multivariate probit model found that mulching has higher prospect to be adopted by organic farmers as it is the most traditional way of soil fertility management practice and organic farmers mainly follow traditional way of farming. It is also adopted by those having higher farm income as it indicates producing more crops which further provides more crop residue for mulching. Those who have taken credit also have higher propensity to adopt mulching, which might not have direct impact, but credit for higher investment in crop cultivation results in higher crop residue for mulching. However, in some instances adoption can be hindered by lack of fund, such as in the case of compost-shed. Thus, it is advisable that fund assistance should be increased so as to increase adoption rate of compost-shed by the majority. Tenant farmers have less resource holding which is why their probability of constructing biogas (that ultimately provides bio-slurry) decreases, as it requires higher initial investment. Similarly, those who have bigger farm size or higher farm income indicate being resource rich, and thus their chances of adopting higher investment requiring practices such as compost-shed and biogas, respectively, too increase. This further proves that financial inability is the major drawback for adoption of these sustainable practices.

One of the ways to increase the adoption rate is training as it complements technical knowledge required to implement these practices. Also if such practices largely rely on locally available resources such as bio-pesticides, then even tenant farmers facing financial constraint can adopt these practices. Farmers also tend to adopt most of such practices in tandem. They are practicing uncommon techniques such as plastic cover and/or vermicompost along with biogas, bio-pesticides, or even with traditional ones like mulching. It indicates that any additional organic means of soil fertility or pest management practices can be introduced to those households who are already adopting one of such practices. But sometimes, these practices become substitutes because of their nature of relying on same input such as mulching and biogas that directly or indirectly depends on crop residue. Thus, any effort to enhance such adoption rate can consider these characteristics of various practices. Hence, adoption of organic means of crop management practices is influenced in different ways by various socioeconomic factors that should be regarded before any intervention.

The third component included under environmental implication of these two farming systems is analyzing crop diversity. Crop diversification benefits environmentally (biodiversity, pest control, resource use efficiency, nutrient cycling processes, resilience, low weed infestation, and nitrate leaching), socially (dietary need, employment opportunities by cultivating crops all year round), and economically (high-value crops). Crop diversity allows resource use efficiency through facilitation and complementarity between species. It maximizes profit; minimizes risk; conserves soil; improves soil fertility; controls weeds, pests, and diseases; and provides balanced nutrition. This study uses Shannon Diversity Index that captures both richness (number) and evenness (abundance) of species. Organic farming in the study areas is richer in integrating more number of crop types (richness) but is poor in evenness, which resulted in having lower Shannon Diversity Index than conventional farming. Since crop evenness is a better indicator of improved productivity than crop richness, it can be implied that farmers, especially organic farmers, should be made aware of this fact. Result from ordinary least square model also showed education attainment, livestock holding, non-farm income, group membership, and training to have significant positive impact on Shannon Diversity Index. Clearly, educated farmers have more knowledge on benefits of having various crops and its benefits to health. Non-farm income allows farmers to grow diverse crops for own household consumption rather than having to specialize for increasing income. Membership in a group formed for the purpose of organic farming and training related to organic farming can improve Shannon Diversity Index because the purpose of such group formation and training is to make farmers aware of benefits of agroecological principles resulting from crop diversity. Finally farther distance to market will encourage farmers to have better Shannon Diversity Index, i.e., to grow variety of crops and in abundant quantity because they will prioritize on being self-sufficient and avoid buying or selling in the market to save transportation cost. Easier access to market leading to low Shannon Diversity Index suggests that market is only favorable for few selected crops, which will encourage farmers for crop specialization. Had there been market opportunities for variety of crops, it could have led to diversifying more crops which is also beneficial for overall production through various environmental services while improving income as well. Therefore, any effort to improve Shannon Diversity Index should consider these characteristics. Most importantly, effort should also be made to cultivate crops more evenly in addition to having numerous types to reap more benefit from environmental point of view, ultimately resulting in higher production.

Economic benefit is probably the most important reason for smallholder farmers to undertake any practice. Lower monetary return is a major bottleneck for large-scale adoption of organic farming. Income from organic farming may be increased through improved yield, reduced cost, and access to premium market. This book analyzed farm income, gross farm cash income, production, and net return for this matter. Income from crops in organic and conventional farming is assessed in an effort to enhance monetary benefit from farming activities, especially from organic farming. Income is divided into total farm output valuation and farm cash income generated from selling in the market and was analyzed through ordinary least square and bivariate probit model. The rationality for analyzing these two issues separately against selected set of farmers' socioeconomic characteristics was justifiable from the result which showed same variables having differing impact on total farm income and farm cash income.

Male-headed households, for example, have higher farm income, but their inclination to market crops is negative. The growing tendency of shift among males and young generations alike in non-farm sector has put a strain on farming activities. There were multiple cases where male members are involved in non-farm sector but at the same time would do farming whenever they could. Since marketing crops take considerable amount of time and effort, it seems that such households limit their agricultural activities for own consumption only. In addition to that, non-farm income is also least likely to be invested in farming activities as its impact on farm income is also negative. Likewise, tenant farmers are encouraged to increase farm income but have lower potential of selling crops in the market and generating income therein because of having to pay crops produced as rent. Livestock holding also increases farm income, but it does not translate into being involved in marketing with an equal intensity as livestock rearing is time-consuming, leaving households with less time for marketing purposes. Household head's age plays a crucial role in increasing farm income, while their education, farming as their primary occupation and access to credit, is important for increasing both farm income and increasing their chances of being engaged in marketing the crops. Having more diversified crops with better evenness increases farm income through improved biological activities. Farm size improves both farm income and likelihood of marketing. Farm income, which also accounts for having high-value crops besides the size of the farm, also contributes in increased income from marketing the crops. Market access, credit facility, and market information definitely have important role to play in marketing the crops. One of the interesting findings from this study is that training related to organic farming has positive impact on farm income, but membership in a group formed for the purpose of organic farming has negative impact on probability of being engaged in marketing the crops. More so, farmers from Phoolbari VDC, where access to premium market exists, have lower farm income but higher farm cash income than farmers from other VDCs. While the latter needs more scrutiny, this book recommends to provide training that compensates for lack of education, increase farm size through merging and collaboration, provide access to credit and market information such as crops' price at which consumers buy, and encourage more diversified farm with more crop varieties and evenness to improve income from farming.

Conventional farmers have higher farm income and earn higher farm cash income than organic farmers because at present, production per hectare, commercialization rate, and price per unit for almost all crops are higher for conventional products. In addition to that, access to premium market is very limited and has not been able to make any significant contribution to organic farmers' income. Since monetary benefit can attract farmers to divert their labor force in farming activities and specifically to boost income pertaining to organic farming, making access to premium market is very imperative. Organic farmers should be linked with potential retailers not just in other cities but within the local area as well where few shops have just been commenced to sell organic or eco-friendly agro-products so that farmers would have more bargaining power over the price and quality check of their products. This study also uncovers the fact that among Phoolbari, Shivanagar, and Mangalpur VDCs, the latter two should be prioritized more for increasing the adoption rate of organic farming or improving farming performance in general because farmers in these two areas have lower organic farming adoption rate, Shannon Diversity Index, and gross farm cash income.

This book also compares production and net return from carrot, potato, and cauliflower cultivation, which according to key informant are among the most commercial non-staple crops and have significance in daily food consumption as well as in national crop production. It is analyzed using t-test and ordinary least square model. From the t-test, the average production was found to be higher for conventional farming for all three crops, of which carrot and potato have significant difference. Average net return from conventional potato is found to be significantly higher. As for carrot, conventional method and for cauliflower, organic method have better net return, although both are not statistically significant. However, the aggregate production of all three crops showed organic production to be higher than conventional production in ordinary least square model. This is in contrast to the overall farm income that was found to be higher in conventional farming, one of the reasons being higher crop production per hectare. This indicates that while the aggregate crop production might be higher for conventional farming, it is not necessarily so on a crop-by-crop basis. In other words, it shows that organic farming does not necessarily result in lower crop production. Male-headed households have better crop production because of better access to resources responsible for improving the production. Larger farm size decreases production implying that resource-rich farmers are less obligated to improve production to sell the excess in the market for higher-income generation. Farther distance to agrovet will have negative impact on crop production because of difficulty in accessing production-enhancing inputs that are available through agrovets. Investment in land under cultivation, seed, and organic inputs such as livestock manure, bio-pesticides, and/or EM; chemical inputs such as chemical fertilizers, pesticides, and micronutrients; and tillage will have positive impact on production. It means that additional unit of these factors will result in higher production. Although investment in chemical fertilizers, pesticides, and/or micronutrients results in higher production, its long-term use is known to deplete soil fertility as has been shown by numerous literatures. Thus, to upgrade the crop production in a more sustainable way, organic inputs should be enhanced.

Global growth of organic market that is mainly concentrated in developed countries has resulted in smallholder farmers in developing countries to face numerous difficulties in the way of lack of adequate financing, management skills, consistency in workforce, logistics, partnership and cooperation, cultural differences, difficulty in obtaining and maintaining internationally recognized standards, high level of record keeping, delay in procuring certification, cost of certification, and annual reinspection cost. Moreover, specialization, capital intensification, export orientation, increased processing, packaging, and long-distance transporting that are controlled by few large corporate retailers have jeopardized the very fundamentals of organic farming. Thus, in such scenario, the importance of local organic market in developing countries can no longer be underestimated, which seems to be trending upward, but since much of it occurs on informal basis, there is no proper information to know its status. Nepalese local organic market also has similar characteristics. Thus, this book also looks into the present status of organic products available in the local market.

The study of local organic market is confined within two major cities, Kathmandu metropolitan and Lalitpur sub-metropolitan, of the country where much of the market for organic products are located because of potential consumers having better awareness of consuming organic products and higher purchasing power to pay for its higher cost. Various outlets, mostly organic and some conventional, were visited in these two adjoining cities for comparison. A total of 15 categories of products were identified: cereals, vegetables, spices, pulses, oil seeds, fruits, tea, coffee, juice, wine, pickles, honey, jam, snacks, and skin care products, of which vegetables have the highest variety. All the products were packaged in some form except for vegetables, spices, and fruits; about less than half of which were sold without the package. All items under pulses, oil seeds, juice, wine, pickles, jam, snacks, and skin care products are self-claimed as organic, which means that they are sold on the basis of word of mouth. For the rest, although most of them are also classified as self-claimed, some are certified by designated certifiers such as National Association for Sustainable Agriculture Certified Organic, Organic Certification Nepal, Société Generale de Surveillance, and Organic Agriculture Certification Thailand, while others are approved by various agencies as safe to consume such as Vegetable Development Directorate, National Tea and Coffee Development Board, International Organization for Standardization ISO-22000:2005, and Department of Food Technology and Quality Control. Some are also found to be merchandised under "fair trade," but sellers use it anyways to market their products as organic. Premium for organic products ranged from 2% to 40%. Surprisingly some self-claimed organic products were found to have higher premium rate than those certified. This can be attributed to the fact that in Nepalese context, some farmers even without certification are able to get premium price purely based on mutual trust or personal links, whereas others are devoid of such benefit despite of being certified because of poor marketing system and skill. On average, premium for cereals is 21%, vegetables is 17%, pulses is 20%, and fruits is 23%. Sellers opined that the popularity of organic products is growing gradually, not just among foreigners but local people as well, although natural calamities and political instability have caused damage to its growth. Complication of certification process adds to the already challenging situation especially for small organic producers and thus should be simplified, made convenient, and inexpensive.

Organic farming relies on locally available resources and techniques such as crop rotation, animal manure, green manure, natural enemies, pest-free plant varieties, companion planting, integrated pest management, etc. to control pests, weeds, and diseases and maintain health of soil and that of all living organisms involved as well. Among the number of available options, it is difficult for farmers to understand which method could be more efficient than the other. Thus, knowing the best choice among many alternatives can be valuable to increase production under a given environmental condition. With this understanding, this study evaluated production of carrot, kidney bean, and potato, which were chosen for their commercial value, share in daily food intake, and importance at national level as well. About ten plots sized $7.5m^2$ each were allocated for each crop among which two plots were allocated for experimenting crop production with and without irrigation; two plots for with and without mulching using straws; two plots for with and without self-made bio-pesticides; and remaining four plots for pruning at the rate 0%, 25%, 50%, and 75%. In all three crops, irrigation gave better production than without irrigation, mulching gave better production than without mulching, and bio-pesticides gave better production than without bio-pesticides. Within pruning, crops without pruning gave better production, and the higher the pruning rate, the less became the production. In all three crops and across all plots, mulching resulted in the highest production, although all crop productions were higher than Chitwan district's average production rate, except for pruning which was significantly lower in all plots allocated for pruning, in case of all three crops.

From this book, it is found that there is an ambiguous dissimilarity between organic and conventional farming systems when measured in terms of all three dimensions of sustainability: social, economic, and environmental. Such could be the situation in other developing countries as well where, as presumed earlier, there is often a vague difference between organic and conventional farming systems in a sense that both incorporate integrated farming of crops and livestock. Socially, organic farming is able to form a social fabric through group formation where farmers can gain knowledge through training and interaction. Environmentally, the soil quality for both farming systems is poor in terms of texture, pH, micronutrients, phosphorus, and potassium level. Higher adoption rate of organic means of crop management practices in organic farming indicates it will be more sustainable overtime. The crop diversity as calculated using Shannon Diversity Index showed that organic farming scored lower than conventional farming. Economically, conventional farming tested better than organic farming as the former has higher farm income and earns higher farm cash income than the latter. While one of the reasons for higher farm income in conventional farming is the higher overall crop production per hectare, the combined production of all three selected crops, carrot, potato, and cauliflower was higher in organic production. This shows that crop-wise, organic production might not always be lower than conventional production. The most established market for organic products within the country exists in Kathmandu and Lalitpur cities where premium ranges from 2% to 40%, but since farmers in Chitwan district have very little means to connect to such market, it further puts strain for organic farmers. Therefore, no one farming system is better in all aspects tested under three sustainability dimensions. However, the concept of time horizon also has a role to play in understanding the concept of sustainability as usually environmental change takes place after many years of accumulated processes. Thus, agro-system that appears sustainable today could be in the process of being unsustainable over a long time period. Various studies have supported the negative consequences brought on by conventional farming and positive aspects of organic farming that takes a much more sustainable approach. Therefore, we should be more careful while glorifying the benefits of conventional farming with only present context in mind.

With the understanding that organic farming is a more sustainable approach toward the future of food security, some of the ways that can help support for its development have been discussed in a broader sense. For the growth and development of organic farming in a developing country like Nepal, there should be pocket areas designated for organic production especially for smallholder farmers. In order to develop such areas, those regions where farming practice is still traditional or is claimed as organic by default or has minimal usage of chemical inputs should be targeted. It will then take minimum effort to make conversion to organic farming than it is to convert conventional farming. Nepal is endowed with ecological diversity which allows to produce different crop varieties in different seasons. Identifying location-specific crop to be grown and technology to be used in different pocket areas where it is most likely to be suitable could be another important strategy. While full conversion of organic farming might not be possible, it is better to focus on specific crops and specific location in different geographical regions of the country which provides a suitable environment for different agricultural products. This shall help divert the needed resources to support production and marketing of crops based on a particular location as well.

Developing such zones not only serves as a basis of minimizing contamination of organic farm from restricted inputs, especially through conventional farm, but also provides various other benefits. To begin with, it makes it easier to form the group that can be a platform to share information and ideas among farmers since organic farming is very much knowledge intensive. In addition, these smallholder farmers when aggregated into a large group can make themselves known among other stakeholders and collaborate for shared benefit. Such as in the case of the study of VDCs, they are able to receive various assistances and extension services from academicians and non-governmental and governmental organizations. Of course there are many challenges when it comes to operating any system with huge number of people with differing beliefs and opinions involved. But it cannot be denied that through such group formation, these farmers are able to get some form of recognition and hence the benefit, which otherwise would have taken humongous resources for a single farmer to achieve. It also increases the probability of their recurring problems to be heard by the concerned actors/institutions.

Besides the group formation, other important socioeconomic characteristics that are found to have significant contribution either in isolation or in combination on the adoption of organic farming or organic farming-friendly practices or technologies, improving crop richness and evenness and/or farm-related income, are training, commercially available organic inputs, education, and livestock holding. Thus, these characteristics should be supported for better return from organic farming. Since most males are engaged in non-farm sector, it is advisable that females be given equal emphasis and supported as they are taking a lead role in decision-making and management related to farming. Financial ability also plays equally important role. Facilitating credit at minimized rates, relief, and subsidy for purposes related to organic farming can add further to lessen farmer's burden.

Premium market probably is the most important motivation for majority of smallholder farmers, if not all. Group formation allows for using the cheapest form of certification such as participatory guarantee system and for the collective selling of organic products. Agribusiness Policy of 2006 has already proposed to establish organic/pesticides-free production area based on commercial crop/commodity production, organic/pesticide-free production, and agricultural product export. These are very much relevant to encourage group conversion and at the same time link them to the potential market based on their expertise of crop production determined by their socioeconomic and ecological context. Marketing of organic products should not only be export oriented, but local market should be given equal importance as farmers will have more bargaining power over the price and quality check of their products. For this, market-related information should be readily available to farmers for effective decision-making. Overall, this calls for a special provision for local- and export-oriented organic market.

Another interesting finding from this study is that eventually farmers have become more aware of the health benefits of organic produces and harmful effect of conventional produces which is why some also tend to grow same crop organically for home consumption and conventionally to sell in the market. Thus, organic farming should be equally promoted for its health benefit rather than just as a profitable endeavor. In the study areas, crop varieties suitable for organic farming are found in limited quantity. In fact many are not even aware of the existence of such varieties because of which they are compelled to use the ones that demand some amount of pesticides. In such scenario, the research and extension institutes that can work directly with such farmers will play an important role. Emphasis on testing and improving soil quality should also be given through training or workshops to the farmers. The study also found that organic farming does not necessarily result in lower crop production and it actually differs from one crop to the other. That is why field experiments that test viability of different inputs and methods in improving organic production, especially through farmers' group in order to engage the end users directly, is also necessary. The documentation to prove that organic farming is beneficial should be well documented and communicated to different agencies. This helps to bring it in the limelight and would encourage more actions from like-minded people. It is known that the government has developed plenty of organic-friendly policies but is weaker on the implementation side. Thus, the government should also be equally enthusiast and active in encouraging this sector by linking farmers, consumers, researchers, traders, certifiers, and other stakeholders.

Appendices

Appendix I: Information on Formal/Informal Groups Formed for the Purpose of Organic Farming

Features of group/VDCs	Phoolbari	Shivanagar	Mangalpur (a)	Mangalpur (b)	Mangalpur (c)
Group type	Cooperative	Informal	Informal	Informal	Informal
Established (year)	2005	2010	2010	2011	2011
Members:					
Male	42	9	1	1	4
Female	83	35	29	29	26
Total	125	44	30	30	30
Farmers Field School (times conducted)	13	6	2	1	1
Certified	Twice	Never	Never	Never	Never
Member saving and loan facility	Yes	Yes	Yes	Yes	Yes

Source: Field survey (2013)

Appendix II: List of Types of Crops Under Six Broad Categories Cultivated in the Study Areas

English	*Nepalese*	*Scientific name*
I. Cereals		
Rice	*Dhan*	*Oryza sativa* L.
Maize	*Makai*	*Zea mays* L.
Wheat	*Gahu*	*Triticum aestivum* L.

(continued)

English	Nepalese	Scientific name
Barley	*Jau*	*Hordeum vulgare* L.
Oat	*Jai*	*Avena sativa* L.
Finger millet	*Kodo*	*Eleusine coracana* (L.) Gaertn.
Common buckwheat	*Mithe phapar*	*Fagopyrum esculentum* Moench
II. Vegetables		
Cauliflower	*Cauli/Fulgobhi*	*Brassica oleracea* var. *botrytis* L.
Cabbage	*Bandagobhi/ Patgobhi*	*Brassica oleracea* var. *capitata* L.
Broccoli	*Brocauli*	*Brassica oleracea* var. *italica* Plenck
Kohlrabi	*Gyathgobhi*	*Brassica oleracea* var. *gongylodes* L.
Tomato	*Golbheda/ Tamatar*	*Lycopersicum esculentum* Mill.
Brinjal/eggplant	*Bhanta/Baigan*	*Solanum melongena* L.
Bitter gourd	*Tito karela*	*Momordica charantia* L.
Lady's finger/okra	*Bhidi/Ramtoriya*	*Abelmoschus esculentus* (L.) Moench
Hot pepper/chilli	*Piro khursani*	*Capsicum frutescens* L.
Scotch bonnet chilli	*Akabare khursani*	*Capsicum chinense* Jacq.
Sweet pepper	*Bhide/Macha khursani*	*Capsicum annuum* L.
Common cucumber	*Asare kakro*	*Cucumis sativus* L.
Vegetable marrow/pumpkin	*Pharsi*	*Cucurbita pepo* var. *medullosa* Alef.
Squash	*Jukini pharsi*	*Cucurbita pepo* L.
Bottle gourd/calabash	*Lauka*	*Lagenaria siceraria* (Molina) Standl.
Sponge gourd	*Ghiraulo*	*Luffa cylindrica* (L.) M. Roem.
Snake gourd/serpent gourd	*Chichindo*	*Trichosanthes anguina* L.
Balsam apple	*Barelo/Barela*	*Momordica balsamina* L.
Chayote/Christophine	*Iskus*	*Sechium edule* (Jacq.) Sw.
Pointed gourd	*Parabar/Parwal*	*Trichosanthes dioica* Roxb.
Ash gourd	*Kubhindo*	*Benincasa hispida* (Thunb.) Cogn.
Watermelon	*Tarbuja/ Kharbuja*	*Citrullus vulgaris* Schrad.
Other Cucurbitaceae:	*Anya phal tarakari:*	
Garden pea	*Matarkosa*	*Pisum sativum* L.
Field bean	*Simi*	*Phaseolus vulgaris* L.
Cowpea	*Bodi*	*Vigna unguiculata* (L.) Walp.
Fava bean/broad bean	*Bakula*	*Vicia faba* L.
Soybean	*Bhatmas/ Bhatmaskosa*	*Glycine max* (L.) Merr.
Other Leguminosae:	*Anya kose tarakari:*	
Mustard greens/leaf mustard/Indian mustard	*Rayo ko saag*	*Brassica juncea* (L.) Czern.

(continued)

English	Nepalese	Scientific name
Garden cress	Chamsur ko saag	Lepidium sativum L.
Spinach	Palungo ko saag	Spinacia oleracea L.
Indian rape	Tori ko saag	Brassica rapa subsp. dichotoma Roxb.
Buckwheat greens	Fapar ko saag	Fagopyrum esculentum Moench
Fenugreek leaves	Methi ko saag	Trigonella foenum-graecum L.
Green garlic	Hariyo lasun	Allium sativum L.
Onion green	Hariyo pyaj	Allium cepa L.
Pumpkin shoot	Farsi ko munta	Cucurbita moschata Duchesne
Colocasia leaf	Karkalo/Gaava (Pidhaalu)	Colocasia esculenta (L.) Schott
Other leafy vegetables:	Aanya saag:	
Radish	Mula	Raphanus sativus L.
Turnip	Salgam/ Gantemula	Brassica rapa L.
Carrot	Gajar	Daucus carota L.
Onion	Pyaj	Allium cepa L.
Garlic	Lasun	Allium sativum L.
Other root vegetables:	Anya jare tarkari:	
Cassava	Tarul	Manihot esculenta Crantz
Colocasia/taro	Pidalu	Colocasia esculenta (L.) Schott
Elephant foot yam	Ole	Amorphophallus paeoniifolius (Dennst.) Nicolson
Sweet potato	Sakhar khanda	Ipomoea batatas (L.) Lam.
Other tuber vegetables:	Anya kandamul:	
Asparagus	Kurilo	Asparagus officinalis L.
Potato	Aalu	Solanum tuberosum L.
	Jhute ghiraula	
Luffa gourd	Pate ghiraula	Luffa acutangula (L.) Roxb.
Winged bean	Pate simi	Psophocarpus tetragonolobus (L.) DC.
Velvet bean	Kause simi	Mucuna pruriens (L.) DC.
Chinese leek	Chinese saag	Allium tuberosum Rottler ex Spreng.
Green amaranth	Latte ko saag	Amaranthus viridis L.
	Pitpite	
III. Spices		
Coriander	Dhaniya	Coriandrum sativum L.
Turmeric	Besar	Curcuma domestica Valeton
Ginger	Aduwa	Zingiber officinale Roscoe
Aniseed	Souf	Pimpinella anisum L.
Bay leaf	Tejpatta	Laurus nobilis L.
Betel nut/areca nut	Supari	Areca catechu L.
	Marathi	
Field mint	Patena/Pudina	Mentha arvensis L.

(continued)

English	Nepalese	Scientific name
Jimbu	*Jimbu*	*Allium hypsistum* Stearn
Fenugreek	*Methi*	*Trigonella foenum-graecum* L.
Chinese parsley	*Chinese dhaniya*	*Coriandrum sativum* L.
	Rose beri	
IV. Pulses		
Kidney bean	*Rajma*	*Phaseolus vulgaris* L.
Black gram	*Kalo mas*	*Vigna mungo* (L.) Hepper
Mung bean	*Mungi mas*	*Vigna radiata* (L.) R. Wilczek
Red gram/pigeon pea	*Rahar*	*Cajanus cajan* (L.) Millsp.
Red lentil	*Musuro*	*Lens culinaris* Medik.
Chickpea	*Chana*	*Cicer arietinum* L.
Garden pea	*Kerau*	*Pisum sativum* L.
Field pea	*Sano kerau*	*Pisum sativum var. arvense* (L.) Poiret
Cowpea	*Bodi*	*Vigna unguiculata* (L.) Walp.
Other common field beans:		
Grass pea/Indian pea	*Khesari*	*Lathyrus sativus* L.
Rice bean	*Masyang*	*Phaseolus calcaratus* Roxb.
Soybean	*Bhatamas*	*Glycine max* (L.) Merr.
Broad bean	*Bakulo simi*	*Vicia faba* L.
V. Oil seeds		
Indian rape/mustard	*Tori*	*Brassica rapa* L.
Mahua seed	*Tora*	*Madhuca longifolia* (J. Koenig ex L.) J.F. Macbr.
Indian colza	*Sarsyun/Sarson*	*Brassica rapa* subsp. *trilocularis* Hanelt
Sunflower	*Suryamukhi*	*Helianthus annuus* L.
Perilla	*Silum*	*Perilla frutescens* (L.) Britton
VI. Fruits		
Guava	*Amba*	*Psidium guajava* L.
Grape	*Angur*	*Vitis vinifera* L.
	Amara	
Pomegranate	*Anar*	*Punica granatum* L.
Mango	*Aap*	*Mangifera indica* L.
Peach	*Aaru*	*Prunus persica* (L.) Batsch
Indian gooseberry	*Amala*	*Phyllanthus emblica* L.
Sugarcane	*Ukhu*	*Saccharum officinarum* L.
Avocado	*Avocado*	*Persea americana* Mill.
Pineapple	*Bhui katahar*	*Ananas comosus* (L.) Merr.
Jackfruit	*Rukh katahar*	*Artocarpus heterophyllus* Lam.
Lemon	*Kagati*	*Citrus limon* (L.) Osbeck
Java plum	*Jamun*	*Syzygium cumini* (L.) Skeels
Black mulberry	*Kimbu*	*Morus nigra* L.
Banana	*Kera*	*Musa* × *paradisiaca* L.
Rose-apple	*Gulab jamun*	*Syzygium jambos* (L.) Alston
Papaya	*Mewa*	*Carica papaya* L.

(continued)

English	*Nepalese*	*Scientific name*
Litchi	*Licchi*	*Nephelium litchi* Cambess.
Kumquat	*Muntala*	*Citrus japonica* Thunb.
Indian plum	*Bayer*	*Oemleria cerasiformis* (Torr. & A.Gray ex Hook. & Arn.) J.W.Landon
Coconut	*Nariwal*	*Cocos nucifera* L.
Common pear	*Naspati*	*Pyrus communis* L.
Bayberry	*Kafal*	*Myrica gale* L.
Pummelo	*Bhogate*	*Citrus grandis* (L.) Osbeck

Appendix III: List of Organic Products Identified in the Local Market

I. Cereals	Bottle gourd	Green salad
Barley (flour)	Brinjal	Lady's finger
Buckwheat (flour)	Broad bean	Lettuce
Millet (flour)	Broccoli	Mixed salad
Millet (white)	Cabbage	Mixed vegetables
Rice (basmati, brown, red, Taichin, white)	Carrot	Mushroom (shiitake)
Wheat (flour)	*Capsicum*	Nettle
II. Vegetables	Cauliflower	Onion
Basil	Chayote	Parsley
Bean	Chilli (Akhabare – green/red, dried)	Potato
Beet root	Coriander	Pumpkin
Bitter gourd	Cucumber	Radish
Bok choy	Garden peas	Rosemary
Spinach	IV. Pulses	I. Drinks
Sponge guard	Beans (mixed, pink, Rice, white)	Tea (24 types)
Sweet potato	Gram (blended, brown bean)	Coffee (41 types)
Tomato	Horse gram	Sabagaard raspberry juice
Turnip	Jumli dal	Sabagaard lemon juice
Onion	Kidney bean	Sabagaard blueberry juice
Pea	Lentil (black, red)	Sabagaard strawberry juice
Potato	Soybean (black, brown)	Grapple juice
Pumpkin	V. Oil seeds	Orange juice
Radish	Flaxseeds	Yacon syrup

(continued)

Spring onion	Organic quinoa seeds	Wine
String bean	VI. Fruits	Probiotic fruit wine
Sweet potato	Almond	Sisnu wine
Spinach	Apple (+dried, sliced)	II. Others
Watercress	Apricot (dried)	Oyster mushroom pickle
Zucchini	Avocado	Fenugreek pickle
III. Spices	Blue berry (dried)	Jackfruit pickle
Bay leaf	*Citrus limetta*	Bitter gourd pickle
Black sesame	Coconut (+dry)	Lemon pickle
Chilli (powder)	Junar orange	Hot chilli pickle
Coriander	Lemon	Honey (27 types)
Cumin	Kiwi	Apricot jam
Fenugreek	Mandarin	Hog plum jam
Garlic	Orange	Ginger jam
Ginger	Papaya	Guava jam
Garam masala	Pear	Papaya jam
Jimbu	Persimmon	Snacks
SHS mix masala	Pineapple (dried)	Hog plum candy
Sichuan pepper	Plum (dried)	Potato chips
Turmeric	Pomegranate	Skin care products
	Strawberry	Organic silk soap
	Walnut	Organic hair and skin care set
	Water melon	Yak milk soap
		Aloe vera gel

Source: Field survey (2015)

Index

A
Academicians, 65
Accessibility, 182
Accreditation, 31
Acid-alkaline, 73
Acidic, 184
Acidification, 84
Acidity, 75
Adaptation, 26, 30
Adoption, 7, 60, 84, 180–182
Adoption rate, 98
Advocacy, 30
Aeration, 73, 75
Afghanistan, 9
Africa, 3, 7, 8
Age, 47, 49, 60, 87, 106, 118, 135, 182, 186
Agriculture, 6, 9, 21, 37, 84, 113, 180
    policies, 180
    practices, 84
    productivity, 23
    technologies, 113
Agro-biodiversity, 27
Agrochemicals, 3, 5, 23, 37, 40, 51, 59, 133, 134, 145, 181
Agro-climate, 21, 180
Agroecology, 6, 30, 133, 183
Agroforestry, 26, 31
Agro-inputs, 110
Agro-products, 125
Agrovets, 49, 50, 61, 92, 107, 118, 139, 182, 183
Alkaline, 75
Ammonium sulfate, 77
Animal manure, 6, 71, 168, 179, 189
Anthropogenic, 4, 26

Anti-feedant, 170
Argentina, 8
Ash, 71
Asia, 3, 8, 151
Asia-Pacific, 103
Asparagus, 39
Assistances, 53, 160, 182, 185
Atmosphere, 5
Australia, 8, 15
Austria, 8
Avocado, 159, 163
Awareness, 41, 180

B
Backward elimination, 119
Bangladesh, 9, 153
Beekeeping areas, 11
Bhutan, 9
Binary logistic model (BLM), 63, 183
Bio-
    fertilizer, 24, 31, 41
    pesticides, 24, 31, 41, 53, 60, 66, 84, 85, 143, 168, 184, 189
    slurry, 85, 184
Biodiversity, 3–5, 30, 83, 103, 179
Biodynamic farming, 6
Biogas, 85
Bio-inputs, 14
Biological, 6, 73, 84
Bivariate probit model (BPM), 120
Black, 163
Bone meal, 78
Boron, 94, 143
Brazil, 153

Breusch-Pagan/Cook-Weisberg, 64, 94, 108, 121, 143
Brown, 161, 163
BT cotton, 13
Buckwheat, 115
Buffer zone, 43

C
Canada, 15
Capital availability, 98
Capital intensification, 152, 188
Carbon dioxide (CO_2), 6, 12, 25
Carbon stocks, 26
Carpentry, 124
Carrot, 115, 134, 136, 168, 187, 189
Cash crops, 14, 39, 168
Cash income, 49, 50
Cauliflower, 134, 136, 187
Central Development Region, 37
Central Russian States, 16
Cereals, 11, 22, 39, 103, 108, 114, 156, 182, 188
Certification, 8, 11, 13, 31, 33, 53, 114, 152, 156, 181, 189
Certification Alliance (CertAll), 15, 156
Certified, 189
Certifiers, 159, 189
Chain shops, 14
Challenges, 114, 126
Charcoal, 71
Chemical, 1, 73, 84
    fertilizers, 3, 23, 27, 28, 32, 38, 43, 61, 66, 84, 94, 143, 179, 183, 184
    inputs, 41, 143, 181
    pesticide, 3, 23, 28, 32, 40, 43, 52, 61, 66, 94, 110, 145, 179
Chicken litter, 142
Chicken manure, 78
China, 7, 21
Chitwan district, 37, 59, 168
Cities
    Kathmandu metropolitan, 153
    Lalitpur sub-metropolitan, 153
Citrus fruit, 11, 39
Clay, 72
Climate, 23
Climate change, 4, 25, 133
Coconut, 163
Coffee, 15, 22, 31, 32, 155–157, 159, 164, 165, 188, 189
Collection center, 125
Collection point, 54
Commercial crops, 123

Commercialization, 22, 49, 50, 61, 118, 153, 180
Commercialization rate, 92, 182, 187
Common Objectives and Requirements of Organic Standards, 12
Community, 4
Community trust, 114
Companies, 14
Companion planting, 168, 179, 189
Compost, 32, 71, 77, 85, 184
Compost-shed, 53, 85, 184
Constraints, 153
Consumers, 4, 14, 134, 153, 160
Consumption, 8, 151
Contamination, 43, 183
Conventional farmers, 44, 89, 104, 184
Conventional farming, 4, 7, 43, 113, 115, 134, 167, 179
Conventional fertilizers and pesticides, 85, 142, 170
Conventional growers, 145
Conversion, 61, 64, 183
Cooperation, 152
Cooperative, 41, 145
Copper, 73
Corn, 104
Corporate retailers, 152
Correlation coefficient, 98
Correlation matrix, 64
Cost components, 140
Cost of, 186
    labor, 142
    land, 142
    micronutrients, 142
    organic fertilizers and pesticides, 142
    post-production, 142
    seed, 142
    tillage, 142
Cost-benefit, 2
Cost-effective, 134
Cotton, 11
Cover crops, 77
Credit, 48, 50, 61, 92, 107, 118, 139, 182, 185, 187
Crops, 181, 184
    cultivation, 185
    diversification, 7, 103, 185
    diversity, 49, 110, 182, 185
    failure, 26, 104, 152
    intensification, 84
    management, 83
    production, 2, 134, 135, 146, 179
    productivity, 71

quality, 125, 170
residue, 98, 184
rotations, 6, 26, 30, 168, 179, 189
specialization, 106, 110
types, 109
varieties, 52, 125, 182
yield, 4, 118
Cultural differences, 152
Cultures, 2

D
Decision-making, 60, 124
Deficit, 41
Degradation, 84
Denmark, 9
Dependent variables, 95, 121
Depletion, 84
Desertification, 3, 179
Desire, 55, 61
Developed countries, 3, 41, 104, 151, 188
Developing countries, 3, 29, 84, 104, 134, 151, 188
Diamond, 41
Diamonium phosphate (DAP), 40, 52, 94, 143
Diseases, 40, 44, 52, 66, 168, 170, 179, 185
Diversified crops, 187
Diversity, 4, 7, 26, 30, 152
Domestic market, 12, 32
Dried pulses, 11
Drought, 30
Dynamic properties, 83, 184

E
Eco-friendly, 13
Eco-friendly agro-products, 187
Ecological, 6
   diversities, 181
   farming, 30
   footprints, 153
   foundations, 4, 5
   richness, 24
Economic, 2, 4, 21, 103, 113, 134, 179
   benefit, 134
   development, 3, 22, 113
   growth, 23, 133
   liberalization, 6
   profitability, 3
   prosperity, 2
   sustainability, 4, 5
   viability, 113
Economies of scale, 4, 7, 60

Ecosystem, 4, 183
Ecosystem services, 179
Education, 4, 47–49, 60, 88, 106, 116, 135, 153, 182, 185, 187
Effective microorganism (EM), 143, 188
Efficiency, 7, 185
Egg, 39
Egypt, 153
Elemental sulfur, 77
Elements, 73
Emission, 6
Employment, 21, 22, 104, 133, 185
Endogeneity test, 121
Energy, 3
Entitlement, 182
Environmental, 2–4, 23, 103, 133, 134, 147, 179
   adversity, 104
   change, 79
   degradation, 2, 3, 26, 179
   health, 27
   integrity, 153
   pollution, 24
   services, 29
   sustainability, 4
Environment-friendly products, 55
Establishments, 123
Europe, 8, 151, 153
European Union, 11, 152
Eutrophication, 78
Evaporation, 170
Evenness, 107, 185
Expenditure, 142
Experience, 49, 60, 87, 106, 118, 135, 140
Exploitative farming, 5
Export, 12, 31, 114, 180, 188
   market, 33, 114
   orientation, 152
Exporters, 24
Exporting, 13
Exporting countries, 152
Exposure, 134
Extension, 33, 180, 181
Extension services, 106, 181
External inputs, 153

F
Fair trade, 151, 159
Falkland Islands (Malvinas), 8
Farm cash income, 127, 182, 187
Farm gate, 50
Farm households, 114, 146

Farm income, 49, 50, 89, 184–187
Farm productivity, 30
Farm size, 49, 56, 61, 89, 106, 118, 135, 182, 185, 187
Farm yard manure (FYM), 40, 52, 66, 84, 168
Farmer's Field School (FFS), 41, 146, 168
Farmers', 4
    knowledge, 26, 84
    resilience, 28
Farming, 1, 13
    behavior, 51
    decision, 60
    systems, 1, 3, 72, 89, 94, 182
Farms, 24
Feather meal, 78
Female-headed, 47, 61
Fertilization, 75, 184
Fertilizer, 2, 31, 74, 118
Field experimentation, 168, 172
Final price, 48, 50, 61, 126
Financial institutions, 55, 182
Fish, 39
Floods, 30
Focal group discussion, 125
Food, 22
    accessibility, 28
    availability, 27
    deficiency, 22, 37
    grains, 5
    insecurity, 26, 84, 179
    production, 1
    production system, 114, 168
    productivity, 23, 29
    residues, 3, 28, 179, 180
    security, 4, 27
    stability, 29
    standards, 31
    surpluses, 3
    utilization, 28
Food-insecure, 3
Forage crops, 78
Forestry, 22
Formalized market, 153
France, 8
Freshwater, 5
Fruits, 15, 22, 108, 114, 156, 188

G
Gender, 48, 118
Genetic engineering, 2
Genetically modified organisms (GMOs), 31, 179
Germany, 8
Global food supply, 135
Global organic market, 151
Global positioning system (GPS), 12
Global sales, 8, 180
Globalization, 8, 152
Government and international organizations, 31, 65
Green manures, 6, 77, 126, 168, 179, 189
Green revolution, 3, 5, 6, 179
Greenhouse gas, 4
Gross domestic product (GDP), 9, 21, 133
Gross farm cash income, 50, 186
Gross margins, 167
Groundwater contamination, 72
Group, 145, 183, 187
    activities, 55
    conversion, 59, 181
    formation, 51–55, 61, 181, 183
    membership, 44, 89, 106, 139, 185
Growth inhibitor, 170
Growth regulators, 179
Guidelines, 181

H
Hausman test, 121
Head of households (HHHs), 47
Health, 4, 23, 51, 133, 147
    benefits, 125
    conscious, 182
    hazards, 40
    implication, 3, 179
Heckman sample selection probit model, 92
Heckman selection model, 119
Herbicide, 94
Heteroscedasticity, 94, 108, 121, 143
Heteroskedasticity, 64
High income, 153
Higher price, 164
High-yielding crop varieties, 2, 179
Hill, 21
Home consumption, 51
Honey, 31, 115, 156, 188
Horizontal diversification, 103
Hormones, 94, 179
Households, 38, 43, 114, 182
    size, 88
    characteristics, 106
Human resources, 14, 22
Hybrid seeds, 143

I
Imbalance of benefits, 3, 4, 179
Importing countries, 152
Incentive, 4

Income, 21, 22, 37, 50, 133, 180
    farm, 114
    gross, 142
    gross farm cash, 114
    source, 48
Independent variables, 95, 121
India, 3, 7, 9, 21, 24, 153
Industrial revolution, 1, 5
Industrialization, 103
Informal farmers' group, 41
Information, 118, 129, 135, 187
Infrastructure, 5, 33, 181
Inherent soil property, 83
Inorganic chemical fertilizers, 71
Input costs, 167
Input-intensive, 23
Inputs, 139
Insecticide, 94, 145
Institutions, 61
Integrated, 22
    crop management, 6
    farming, 181
    pest management, 6, 31, 168, 179, 189
Integrated soil fertility management (ISFM), 85
Integrated soil nutrient management, 31
Intensification, 6
Intensive, 23
Interaction, 109
Intercropping, 103, 152
Interest rate, 53
Internal Control System (ICS), 32
International Federation of Organic Agriculture Movements (IFOAM), 8
International market, 12, 23, 31
Investment, 98, 126, 134, 143, 180, 185
Iron sulfate, 77
Irregular rainfall, 30
Irrigation, 1, 5, 31, 50, 75, 123, 168, 169, 184, 189

J
Jam, 156, 188
Japan, 15, 24, 151
Juice, 156, 188
Justification, 113

K
Kathmandu metropolitan, 153, 188
Kathmandu valley, 27
Kidney bean, 115, 168, 189
Knowledge exchange, 33

L
Labor, 1, 10, 30, 49, 60, 88, 118, 142, 180, 183
    availability, 106, 118, 139, 181, 182
    force unit (LFU), 47
    scarcity, 22
Lalitpur districts, 153
Lalitpur sub-metropolitan cities, 153, 188
Land, 142
Latin America, 3, 8
Legislation, 33, 181
Legume crops, 24, 103, 184
Leguminous cover crops, 135
Lemongrass oil, 15
Liechtenstein, 8
Lime, 71, 74, 77, 184
Limestone, 77
Linkage, 180
Livelihood, 4, 9, 28, 30, 50, 60, 65
Livestock, 22, 49, 118, 181
    feed additives, 179
    holding, 48, 61, 89, 106, 135, 182, 185, 186
    rearing, 123
Livestock unit (LSU), 47
Local, 114
    knowledge, 28
    market, 114, 153, 180
    organic market, 153, 188
    resources, 23
    varieties, 32, 143
Locally available resources, 7, 28, 168, 179, 185
Logistics, 152
Low external input sustainable agriculture, 6
Low input agriculture, 6
Low input sustainable agriculture, 6
Lower middle-income, 9
Lower production, 122, 146
Lower-income, 9
Luxemburg, 9

M
Magnesium deficiency, 78
Maize, 115
Maldives, 9
Male-headed households, 89, 122, 186
Management practices, 66, 84, 152
Mangalpur, 41, 44, 181
Manpower, 33, 181
Manure, 1, 143, 184
Marginal effect, 63, 120
Marginal impact, 147

Market, 4, 41, 49, 50, 61, 85, 92, 106, 114, 118, 133, 134, 139, 182, 187
   accessibility, 28
   failure, 153
   local, 50
   information, 61, 153
   mechanism, 28
   premium, 53
Marketing, 30, 60
   crops, 186
   system, 114, 160, 189
Mass cultivation, 126
Mean, 47
Meat, 39
Medicinal herbs, 32
Meghalaya, 13
Members, 43, 55
   farmers, 45, 52
   participation, 182
Membership, 43, 47, 48, 118, 182, 187
Metabolic disease, 78
Methane, 6, 25
Mexico, 8
Micronutrients, 29, 73, 94, 142, 143, 184
Microorganisms, 73
Middle East, 15
Milk, 39
Minerals, 78
Mitigate, 7, 180
Mitigation, 26, 30
Mixed cropping, 30
Modern inputs, 84
Modernization, 23
Moisture, 73, 85
Monetary benefit, 4, 51, 113
Mono-cropping, 2, 23, 103, 181
Mountain, 21
Mulching, 66, 77, 85, 168, 184, 189
Multicollinearity, 64, 94, 108, 121, 143
Multi-cropping, 110
Multinomial logit (MNL) model, 92
Multivariate probit (MVP) model, 93, 184
Muriate of potash (MOP), 52, 94, 143
Mutual trust, 114, 160, 189

N

National guidelines, 31
National Organic Program (NOP), 11
National Programme for Organic Production (NPOP), 11
National regulation, 14

National standard, 181
Natural enemies, 168, 179, 189
Natural resources, 22, 29, 180
Nepal, 9, 21, 37, 84, 113, 133, 153
Nepal Agricultural Research Council, 74
Net return, 142, 186
Neutrality, 75
Niche organic market, 15, 160
Nitrate leaching, 103, 185
Nitrogen leaching, 27, 71, 72, 183
Nitrous oxide, 6, 25
Non-farm income, 49, 50, 61, 89, 106, 118, 182, 185, 186
Non-farm sector, 124, 146
Non-governmental organization (NGO), 14, 65, 98
Nonmembers, 43, 55
Nontoxic, 170
North America, 8, 151, 153
Nutrients, 73, 75, 78, 83
   budgeting, 78, 184
   cycling processes, 103, 152, 185
Nutrition, 185

O

Oceania, 8
Off-farm income, 139
Oil seeds, 11, 39, 108, 114, 156, 188
Opportunities, 114, 153
Opportunity cost, 125
Ordinary least square (OLS) model, 107, 119, 142, 185
Organic
   amendments, 74
   biomass, 73
   by default, 24, 156, 181
   farmers, 44, 84, 89, 104, 184
   farming, 6, 30, 37, 59, 113, 115, 133, 153, 167, 179
   fertilizers, 31, 33, 78, 142
   growers, 145
   market, 188
   matter, 26, 72, 84, 183
   movement, 153
   outlets, 153, 155
   pesticides, 142
   production, 127
   products, 12, 152–154, 181, 182
Organic Certification Nepal (OCN), 53
Organic Means of Crop Management Practices (OCMPs), 84

Organically feasible varieties, 182
Organics International, 12
Outlets, 188
Own land, 89
Ownership, 48

P
Packaged, 156
Packaging, 145, 152, 159, 188
Pakistan, 9
Partial funding, 98
Participants, 170
Participatory Guarantee Systems (PGS), 12, 32
Partnerships, 31, 152
Patriarchal, 122
Peat moss, 77
Percentage, 47
Perception, 183
Permaculture, 6, 30
Personal links, 160
Peru, 7
Pests, 40, 44, 52, 66, 168, 170, 179, 185
  control, 185
  -free plant varieties, 168, 179
  management, 104, 184
  resistance, 24
Pesticides, 3, 4, 41, 152, 184
pH, 72, 183
Phoolbari VDC, 41, 44, 61, 92, 106, 118, 168, 181
Phosphate, 78
Phosphorus, 27, 71, 72, 183
Physical properties, 73, 84
Pickles, 156, 188
Plain, 21
Plant growth, 73, 77, 170
Plant nutrients, 73
Plant roots, 85
Plant size, 170
Plant varieties, 189
Plastic cover, 85, 184
Plot, 169
Policies, 2, 4, 31
Policy formulation, 31
Political, 2
Population, 3, 4, 21, 26, 37, 135
Postharvest, 8, 14
Postproduction, 142
Potash, 40, 78
Potassium, 27, 32, 72, 183
Potato, 134, 136, 168, 182, 187, 189
Poverty, 14

Poverty alleviation, 22
Practices, 83
Precipitation, 25
Premium, 7, 54, 114, 115, 127, 145, 152
  markets, 51, 115, 126, 182, 186, 187
  price, 28, 114, 134, 153, 182, 189
  rate, 160, 189
Price, 145
  fluctuation, 153
  monopoly, 153
  per unit, 187
Primary macronutrients, 73
Primary occupation, 47, 48, 60, 89, 106, 118, 139, 187
Private organizations, 14, 126
Probability, 67
Procedure, 181
Processing, 13, 32, 145, 152, 181, 188
Processors, 24
Producers, 11, 24, 164, 180
Product quality, 103
Production, 4, 31, 32, 44, 104, 134, 145, 151, 167, 169, 181, 186
  efficiency, 2
  maximization, 104
  per hectare, 187
  risks, 26
Productivity, 4, 5, 22, 27, 30, 38, 84, 114, 153, 179, 185
Professional hazard, 3, 28, 179
Profit, 185
Profit orientation, 113
Profitability, 114
Promotion, 30, 31
Protein, 29
Pruning, 168, 189
Pulses, 39, 108, 114, 156, 169, 188
Purchasing power, 180, 188

Q
Quality, 164

R
Record keeping, 152
Red rice, 14
Regional imbalance, 179
Regression, 64
Regulations, 11, 32, 181
Reinspection, 152
Remunerative job, 122
Renewable energy, 4, 5

Rent, 143
Repellant, 170
Research, 30, 153, 180, 181
Research and extension, 22
Research Institute of Organic Agriculture (FiBL), 8
Residue, 7, 134
Residue-free, 29
Resilience, 30, 185
Resilient, 7, 23, 180
Resilient farming system, 29
Resistance, 26
Resource, 41, 185
    degradation, 4, 179
    endowment, 98
    poor, 98
    use efficiency, 103
Respondents, 45
Retail, 12
Retailers, 182, 187
Retailing, 13
Rice, 23, 115, 182
Richness, 107, 185
Risk, 65, 185
Risk averse, 61
Robust standard errors, 64, 94, 108, 121, 143
Root activity, 74
Root support, 83
Rules, 181

S

Safety, 4
Safety net, 61
Sand, 72
Sandy loam, 75, 183
Savings, 182
Scientific knowledge, 6, 183
Scientific principles, 5
Secondary macronutrients, 73
Seed, 118, 142
Seed germination, 77
Self-claimed, 156, 189
Self-consumed, 114
Self-consumption, 67
Self-reliant, 25, 181
Self-sufficient, 3, 153
Selling crops, 48, 50, 182
Shannon diversity index (SHDI), 50, 107, 118

Shivanagar VDCs, 41, 44, 181
Sikkim, 13
Silt, 72
Skills, 114, 118, 140, 160, 189
Skin care products, 156, 188
Smallholder farmers, 43, 113, 152, 188
Smallholders, 22, 50
Snacks, 156, 188
Social, 2, 4, 103, 134, 179
    and environmental benefits, 113
    network, 32, 61
    sustainability, 4
Socioeconomic, 3, 47, 60, 84, 104, 107, 182
    characteristics, 114
    factors, 119, 134
    shocks, 104
Soft-rock phosphate, 78
Soil, 4, 5, 71, 183
    aggregation, 72, 183
    amendments, 71, 78
    bulk density, 72, 183
    conservation, 85
    degradation, 3
    ecosystem, 72
    erosion, 3, 84, 179
    fertility, 1, 7, 23, 26, 51, 71–73, 83, 84, 133, 146, 184, 185
    health and fertility, 85
    moisture, 85
    organic matter, 26
    and pest management practices, 83
    properties, 72, 73
    quality, 3, 27, 143
    salinity, 72, 183
    stability, 30
    structure, 72, 183
    testing, 72
    texture, 72, 183
Soil texture triangle (STT), 74, 183
Soil type, 74
South Asia, 9, 21, 153, 180
South Korea, 24
Specialization, 106, 152, 188
Species
    evenness, 107
    diversity, 107
    richness, 107
Spices, 15, 22, 39, 108, 114, 156, 188
Sri Lanka, 9, 153

Stakeholders, 65, 126
Standards, 13, 152
Standard deviation, 47
Standardization, 33, 181
Straws, 170, 189
Sub-Saharan Africa, 5
Subsidies, 22, 33, 180
Subsistence farming, 2, 22, 31, 180
Subtropical climate, 22, 38, 180
Subtropical fruits, 11
Supermarkets, 15
Supply chain, 14, 164
Surplus, 41
Sustainability, 2, 6, 52, 79, 179, 182
Sustainable, 23, 133, 168
    farming, 5
    pest management technologies, 85
    rice intensification, 31
Switzerland, 9

T
Tarai, 21, 24, 37
Tea, 15, 31, 156, 188
Teaching, 124
Technical facility, 14
Technical knowledge, 183, 185
Technological innovations, 2
Technologies, 5, 13, 22, 30, 41, 84, 139, 180
Temperate, 11
Temperature, 25, 30, 170
Tenancy status, 118
Tenant, 185, 186
Tenant farmers, 98
Third legal entity, 156
Tillage, 142
Time horizon, 79
Timor-Leste, 9
Toxicants, 170
TraceNet, 12
Trade, 8
Traditional, 2, 98, 183, 184
    farming, 1, 23, 41
    integrated farming, 110
    practice, 99
Trainings, 31, 41, 47–49, 61, 106, 118, 139, 168, 181–183, 185, 187
Transaction costs, 67
Transition, 134
Transport, 5
Transport cost, 153
Transportation, 125
Transporting, 145, 152, 188
Tropical, 11
    countries, 152
    fruit, 39
Trust, 32
Trustworthy, 126
T-test, 75, 142, 183

U
Uganda, 8, 153
Unequal advantage, 53
Unequal distribution, 53
Unequal participation, 53
The United Kingdom, 14
The United States, 8
Univariate probit model, 92
Universal, 5
Unsustainable, 79
Upper middle-income, 9
Urban areas, 114, 153
Urea, 40, 94, 143
US Department of Agriculture, 11

V
Variation inflation factor (VIF), 64, 94, 108, 121, 143
Vegetables, 11, 22, 31, 39, 108, 114, 156, 168, 182, 184, 188
Vermicompost, 66, 85, 168, 184
Vertical diversification, 103
Village development committees (VDCs), 33, 41, 48, 181
Visual attraction, 125
Vitamin, 94, 143
Vulnerability, 152

W
Water, 83
    eutrophication, 179
    holding capacity, 73, 75
    infiltration, 72, 75, 183
    -use efficiency, 30
Weather dependent, 22
Weeds, 44, 168, 179, 185
    growth, 85, 170
    infestation, 103
Weedicide, 94, 145
Wheat, 115
White's test, 64, 94, 108, 121, 143

Wild collection, 11, 14
Willingness to pay, 153
Wine, 156, 188
Wood ash, 77, 184
Wood residuals, 77
Wool, 39
Word of mouth, 156, 189
Workforce, 152
World Trade Organization (WTO), 16

Y
Yield, 134, 135, 186
   losses, 134
   stagnation, 3

Z
Zinc, 94, 143